FOREWORD

The Programme centres on road and road transport research, while taking into account the impacts of intermodal aspects on the road transport system as a whole. It is geared towards a technico-economic approach to solving key road transport issues identified by Member countries. The Programme has two main fields of activity:

- the international co-operation in road and road transport research to provide scientific support for decisions by Member governments and international governmental organisations, and to assess future strategies concerning roads and road transport problems and the priority policy concerns of Member countries;
- the information and documentation programme (IRRD - International Road Research Documentation), a co-operative scheme that provides a mechanism for the systematic world-wide exchange of information on scientific literature and current research programmes.

The scientific and technical activities concern:

- the assessment of urban and inter-urban road transport strategies;
- the development and management of road traffic control and driver communication systems to enhance network efficiency and quality of service;
- the formulation and evaluation of integrated road and traffic safety programmes;
- the construction, preservation and rehabilitation of road infrastructure.

ABSTRACT

The report is the outcome of a study undertaken by a group of scientific experts and bridge administrators in the framework of long-standing co-operation among OECD Member countries to assess strategies and techniques for more effective bridge management. The study is a follow-up to preceding activities on Bridge Inspection (1976), Evaluation of Load Carrying Capacity of Bridges (1979), and Bridge Maintenance (1981). The report begins with a review of current rehabilitation and strengthening policies, with special reference to the financial and economic aspects and to experience gained with the use of comprehensive inspection systems in Member countries. It then describes, in detail, the state-of-the-art of repair and strengthening techniques. Perspectives for future rehabilitation and strengthening policies are outlined, and the concepts and tools for developing a general policy are presented. The report ends with recommendations for research and the Group's general conclusions. The study highlights the need for sustained efforts to preserve the existing bridge stock, and develops broad policy guidelines for bridge rehabilitation and strengthening.

PREFACE

The development of an economic bridge management policy is of vital importance to all OECD Member countries. The sheer scale of new road and motorway construction programmes in the sixties and seventies has masked the increasing load to maintain and rehabilitate the existing bridge stock.

With the general slackening investment in new infrastructure, priority must now be devoted to preserving and efficiently managing the existing bridge stock. The scale of the problem may be gauged from the simple fact that there are an estimated one million bridges with spans greater than 5m in the OECD countries. Many of these bridges are well past the age when ordinary maintenance procedures will suffice to preserve their useful life or to adapt them to the more strenuous demands of present traffic.

Failure to rehabilitate or strengthen existing structures would increase traffic safety hazards and constitute a missed opportunity to make investments with an often unusually high rate of economic return. Poor maintenance and rehabilitation over a period of years will rapidly result in too great a share of resources going on "stop-gap" measures and hence in wasted funds. Intervention thresholds may be lowered during a difficult year, but if rehabilitation and/or strengthening are postponed overlong, makeshift emergency repairs necessitated on safety grounds will soon prove to be ineffective. Large-scale replacement programmes will eventually be called for at a cost completely unrelated to the benefits obtained.

The report describes national rehabilitation and bridge improvement schemes and sets out how allowance is currently made for economic criteria. It reviews the most prominent repair and strengthening techniques emphasizing those that have been recently developed and highlighting the areas where future research efforts are needed. The study report proposes overall bridge rehabilitation and strengthening guidelines based on forecasts of future needs and technico-economic criteria.

TABLE OF CONTENTS

Chapter I

INTRODUCTION 9

I.1 Place of Study in the Framework of Bridge Management 9
I.2 Terms of Reference 9
I.3 Objectives and Scope 10
I.4 Statement of Problem 11
    I.4.1 Maintenance, repair, strengthening 11
    I.4.2 Damage to bridges 12
    I.4.3 Adapting to needs 12
I.5 Special Features of the Field Under Study 12
    I.5.1 A new and pressing problem 12
    I.5.2 Scale of the problem 14
    I.5.3 Difficulty of the economic approach 14
    I.5.4 Technical difficulties 14
    I.5.5 Need to consider the global feature of the problem .. 14
I.6 Structure of Report 15

References 16

Chapter II

REHABILITATION AND STRENGTHENING POLICIES: PRESENT SITUATION 17

II.1 Scope of the Situation 17
    II.1.1 Background 17
    II.1.2 Inspection procedures 19
    II.1.3 Overall review 22
II.2 Financial and Economic Aspects 25
    II.2.1 Direct costs borne by the bridge authority 25
    II.2.2 Basic considerations when assessing future costs ... 26
    II.2.3 Lifespan of bridge elements 26
    II.2.4 Bases for calculating replacement costs 28
    II.2.5 Replacement costs of bridge elements 29
    II.2.6 Consequences of appropriation ceilings 30
    II.2.7 Desirable funding 30
    II.2.8 The United States programme 30
II.3 Needs Associated with Normal Traffic 31
    II.3.1 Adaptation of bridge structures 31
    II.3.2 Strengthening 32
II.4 Effects of Exceptional Convoys 32

Chapter III

REPAIR AND STRENGTHENING TECHNIQUES: PRESENT SITUATION 33

- III.1 Introduction 33
- III.2 Condition Evaluation (Structures and Materials) 35
  - III.2.1 Results of inspection 35
  - III.2.2 Assessment and special tests 35
  - III.2.3 Guide for concrete defects 40
  - III.2.4 Underwater inspection 41
- III.3 Repair of Foundations 42
  - III.3.1 Repair methods 42
  - III.3.2 Examples of repair work in Member countries 46
  - III.3.3 Splicing new steel sections to H-piles under or above water 48
- III.4 Masonry Structures 49
  - III.4.1 General 49
  - III.4.2 Foundations 50
  - III.4.3 Arches 50
  - III.4.4 Widening 51
- III.5 Concrete Structures 51
  - III.5.1 Scope of the problem 51
  - III.5.2 Repair and strengthening strategy 52
  - III.5.3 Chloride contaminated concrete 52
  - III.5.4 Cathodic protection of reinforcing steel in concrete decks 52
  - III.5.5 Replacement of corroded reinforcement steel 54
  - III.5.6 External reinforcement 54
  - III.5.7 Supplementary prestressing 55
- III.6 Iron and Steel Structures 60
  - III.6.1 Guide for defects in steel structures 60
  - III.6.2 Iron bridges 61
  - III.6.3 Steel bridges 61
  - III.6.4 Suspension and cable stayed bridges 64
- III.7 Composite Structures and Moveable Bridges 64
  - III.7.1 Composite structures 64
  - III.7.2 Moveable bridges 65
- III.8 Components and Accessories 65
- III.9 Seismic Strengthening (Retrofitting) 65
  - III.9.1 Bearings and expansion joints 66
  - III.9.2 Columns, piers and footings 66
  - III.9.3 Abutments 67
  - III.9.4 Liquefaction of foundation soil 67
- III.10 Structural Analysis and Load Testing 68
- References 69

Chapter IV

REHABILITATION AND STRENGTHENING POLICIES: PROPOSALS

- IV.1 The Future Challenges 70
  - IV.1.1 The need for prediction 70
  - IV.1.2 Future traffic demands 70
  - IV.1.3 Environmental demands and considerations 71
  - IV.1.4 Technical restrictions 72

ROAD TRANSPORT RESEARCH

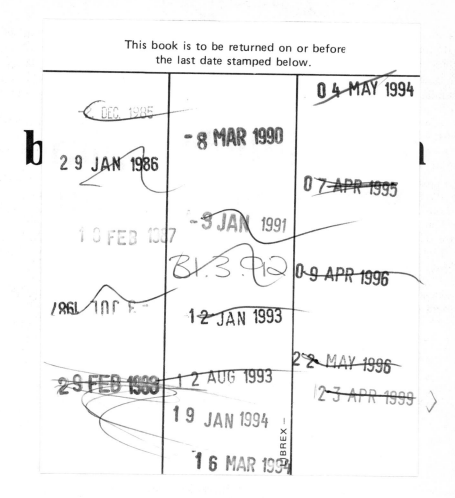

ORGANISATION FOR ECONOMIC CO-OPERATION AND DEVELOPMENT

Pursuant to article 1 of the Convention signed in Paris on 14th December, 1960, and which came into force on 30th September, 1961, the Organisation for Economic Co-operation and Development (OECD) shall promote policies designed:

- to achieve the highest sustainable economic growth and employment and a rising standard of living in Member countries, while maintaining financial stability, and thus to contribute to the development of the world economy;
- to contribute to sound economic expansion in Member as well as non-member countries in the process of economic development; and
- to contribute to the expansion of world trade on a multilateral, non-discriminatory basis in accordance with international obligations.

The Signatories of the Convention on the OECD are Austria, Belgium, Canada, Denmark, France, the Federal Republic of Germany, Greece, Iceland, Ireland, Italy, Luxembourg, the Netherlands, Norway, Portugal, Spain, Sweden, Switzerland, Turkey, the United Kingdom and the United States. The following countries acceded subsequently to this Convention (the dates are those on which the instruments of accession were deposited): Japan (28th April, 1964), Finland (28th January, 1969), Australia (7th June, 1971) and New Zealand (29th May, 1973).

The Socialist Federal Republic of Yugoslavia takes part in certain work of the OECD (agreement of 28th October, 1961).

Publié en français sous le titre :

REMISE EN ÉTAT ET RENFORCEMENT
DES OUVRAGES D'ART

© OECD, 1983
Application for permission to reproduce or translate
all or part of this publication should be made to:
Director of Information, OECD
2, rue André-Pascal, 75775 PARIS CEDEX 16, France.

    IV.1.5 Socio-economic aspects 72
    IV.1.6 The future role of strengthening and
       rehabilitation 72
  IV.2 The Formation of a General Policy 73
    IV.2.1 Aim of a general policy 73
    IV.2.2 A division into policy factors 73
    IV.2.3 A management tool 74
  IV.3 Policy Parameters 76
    IV.3.1 Future increase in functional demands 76
    IV.3.2 Environmental impact 78
    IV.3.3 Description of possible alterations 78
    IV.3.4 Considerations regarding rehabilitation
       and strengthening versus replacement 79
  IV.4 Adapting Structures to Emerging Needs 80
    IV.4.1 Evaluation of bridge structures 80
    IV.4.2 Revised design criteria 81
    IV.4.3 Adaptability to rehabilitation,
       strengthening and other improvements 82

Chapter V

RECOMMENDATIONS FOR RESEARCH 83

V.1 Assessment 83
  V.1.1 Traffic loading characteristics 83
  V.1.2 Wind loading characteristics 84
  V.1.3 Load testing 84
  V.1.4 Remnant life 84
  V.1.5 Condition surveys 85
  V.1.6 Analytical assessment 85
V.2 Repairs 86
  V.2.1 Defective concrete 86
  V.2.2 Corroded reinforcement 87
  V.2.3 Fatigue repair 87
  V.2.4 Bridge management 87
V.3 Strengthening 88
  V.3.1 Arch bridges 88
  V.3.2 Old steel structures 88
  V.3.3 Fatigue 88
  V.3.4 Post-tensioning 88
  V.3.5 Bonded plating 89
  V.3.6 Earthquakes 89
V.4 Replacement of Components 89
  V.4.1 Structural elements 89
  V.4.2 Decks 90
  V.4.3 Piers 90
  V.4.4 Cables 90
  V.4.5 Expansion joints and bearings 90
  V.4.6 Waterproof membranes 91
References 91

Chapter VI

CONCLUSIONS AND RECOMMENDATIONS 92

VI.1 The Need for a Policy 92
VI.2 Policy Guidelines 92
  VI.2.1 Needs 92

```
         VI.2.2  The tools ................................................. 93
         VI.2.3  Comparison between rehabilitation
                 (or strengthening) and replacement ................ 93
         VI.2.4  Policy considerations ............................... 93
  VI.3 Performance of Bridges over Time ............................ 94
         VI.3.1  Replacement rate ......................................... 94
         VI.3.2  Knowledge of the existing bridge
                 stock ........................................................ 94
         VI.3.3  Measures to be taken at the
                 construction stage ...................................... 94
  VI.4 Future Needs ......................................................... 95
         VI.4.1  Trends in needs of normal traffic ................. 95
         VI.4.2  Exceptional heavy vehicles ......................... 95
  VI.5 Techniques ............................................................ 95
  VI.6 Resources .............................................................. 96
         VI.6.1  Financial resources .................................... 96
         VI.6.2  Manpower resources ................................. 96
  VI.7 General Conclusions .............................................. 97
```

ANNEX: RESULTS OF ENQUIRY ON SUBJECTS COVERED IN CHAPTER IV .. 98

LIST OF MEMBERS OF THE GROUP 102

Chapter I

INTRODUCTION

I.1 PLACE OF THE STUDY IN THE FRAMEWORK OF BRIDGE MANAGEMENT

Bridges are key elements of the road network. The inadequacy or failure of only one bridge may limit or severely obstruct road traffic over a large portion of the network. For this reason there is clearly a need for a comprehensive policy covering all aspects of bridge management from the opening of the bridge to its replacement, including:

1. Inspection and documentation
2. Maintenance
3. Evaluation of load-carrying capacity
4. Rehabilitation, repair and strengthening
5. Replacement and reconstruction

The first three issues have already been the subject of study by OECD Research Groups (1, 2, 3)(*); the time has now come to look into the fourth as part of this international research effort focussing on the existing bridge stock.

Obviously the different aspects of bridge rehabilitation and strengthening are closely bound up with the topics that have been studied in the previous reports:

- it is only through inspection and evaluation of the load-carrying capacity of bridges along a given route that maintenance or repair and strengthening needs can be identified;
- maintenance and repair are closely linked, as discussed in more detail later.

Choices, too, have always to be made: between rehabilitation and repair on the one hand, and between replacement and reconstruction on the other. Generally speaking, but particularly from the economic standpoint, consideration must be given to the whole range of management aspects, whether it be for defining a general policy or for resolving a specific case.

I.2 TERMS OF REFERENCE

The terms of reference of the Group entrusted with drawing up this report were the following:

---

*) See list of references at the end of this Chapter.

"Background

The increasing number of bridges requiring rehabilitation and/or strengthening is a common feature in many countries and leads to serious problems such as:

- loss of traffic safety and, in certain cases, reduction of structural safety with the resulting necessity for load limitation;
- expenditure of large sums for a premature bridge replacement which could be avoided if adequate funds were devoted to a timely rehabilitation and/or strengthening;
- preservation of those bridges which form part of the historic and cultural heritage of a country.

Intensified bridge inspection has revealed the rapid rate of deterioration on many bridges and the increasing rate of medium to heavy commercial traffic, in particular, on the secondary road system along which the majority of older bridges are located has highlighted the need for establishing a more systematic bridge rehabilitation and strengthening policy.

Tasks of the Group

Against this background, the Group should prepare a report on the following topics:

1. Comparison of rehabilitation and strengthening methods and techniques as used at the present time in different countries. Indication of the possible consequences of a lack of timely rehabilitation and/or strengthening as regards the overall safety aspect and financial implication.
2. Inventory and appraisal of different methods and techniques of bridge rehabilitation and strengthening of structures and components built and constructed by techniques not being used any more (e.g. masonry).
3. Inventory and appraisal of different methods and techniques of bridge rehabilitation and strengthening of structures and components built and constructed by techniques and with material still being used (e.g. reinforced or prestressed concrete).
4. Definition of the goals of a general bridge rehabilitation and strengthening policy, particularly as regards bridge replacement as another alternative. Indication of means to be implemented.
5. Definition of research to be carried out in order to improve bridge rehabilitation and strengthening.

The Group's study shall take into account earlier OECD work on "Bridge Inspection", "Evaluation of Load Carrying Capacity of Existing Bridges" and "Bridge Maintenance". The study shall deal with all normal type bridges; the special case of large bridges shall be excluded."

I.3 OBJECTIVES AND SCOPE

Given the situation described in the first part of the Group's terms of reference, many countries have already had to repair or strengthen road bridges and have acquired special expertise in this area. A large number of specific cases have been tackled and some countries have begun to initiate a general policy of bridge rehabilitation or even, in some cases, bridge replacement.

The Working Group set itself the task of:

- compiling expertise so far acquired in Member countries;
- establishing broad policy proposals for bridge rehabilitation and proposing criteria that would be helpful for Member countries when specifying national policies;
- making technical recommendations in the light of past success and failures;
- assessing the financial implications of the recommendations;
- identifying future research needs.

The Working Group's study is concerned with the rehabilitation, strengthening or modification of bridges so as to bring them in line with actual requirements. It excludes the question of maintenance which has been tackled in the OECD report "Bridge Maintenance" (3). However, it touches on total bridge replacement to the extent that it constitutes an alternative solution to repair or modification, from the angle of either bridge rehabilitation policy for the network as a whole or policy decisions in regard to particular cases.

From the technical standpoint the study does not discuss the very specific problems posed by bridges of an exceptional character or by structures of a very special type (e.g. floating bridges), or by parts of bridges which involve other than civil engineering techniques (e.g. the mechanical parts of moveable bridges). Very small bridges (with a span of less than 2 metres) are excluded.

I.4 STATEMENT OF PROBLEM

I.4.1 Maintenance, repair, strengthening

The road network is crucial for the economic life of a nation. Its permanent serviceability must be ensured, maintaining acceptable conditions of safety for the various vehicular traffic categories for which the network was designed. This applies to the total network, bridges clearly constituting a special feature of the highway structure.

The upkeep or rehabilitation of the bridges in a network form part of one and the same global policy which encompasses maintenance, repair and strengthening.

Maintenance can be considered as the technical aspect of the upkeep of bridges; it is preventative in nature. Repair is the technical aspect of rehabilitation; which is remedial in nature. This distinction may at times seem arbitrary inasmuch as certain techniques and processes serve both purposes. For this reason the report makes extensive reference to the previous study (3) in the matter of techniques. However, in the case of bridges which have deteriorated over the years, the scope of repair is generally considerably larger than that of the usual maintenance operation. The distinction between rehabilitation and strengthening is linked to the aims of the operations in question rather than the techniques that are applied:

- Rehabilitation consists in restoring the bridge to the service level it once had and has now lost. In some cases this consists in giving the bridge the service level which was intended but which has never been attained, because of deficiencies in design or construction. In all cases the purpose of the repair is to remedy these difficulties.

- **Strengthening** consists in endowing the bridge with a service level higher than that initially planned. This may take the form of load-strengthening in the strict sense of the term when the load-carrying structure itself is involved. The term improvement is more appropriate when the operation serves to raise the service level in respects other than load-carrying capacity (e.g. geometric changes of the bridge).

I.4.2 Damage to bridges

A road bridge must be repaired when it can no longer ensure the desirable level of traffic safety or when the maintenance bill is likely to be inacceptably high. This condition may derive from natural ageing, a process which is hastened by lack of maintenance, natural phenomena (floods, earthquakes, etc.) or accidents (e.g. impact of vehicles or ships).

It is only through effective inspection that the need to rehabilitate a defective bridge can be identified [cf. OECD report on "Bridge Inspection (1)].

I.4.3 Adapting to needs

The increase in traffic, both in terms of volume and load, as well as greater requirements on the part of road users and socio-economic constraints are all factors which may necessitate bridge improvement.

The increase in traffic loads may require a corresponding increase in load-carrying capacity which must be ascertained [cf. OECD report on "Evaluation of load-carrying capacity of bridges" (2)]; this involves strengthening.

The need to adapt to actual vehical dimensions may call for improvements in bridge clearance.

Growth of total traffic carried by the bridge may require widening; it may prove necessary to add new load-carrying elements or strengthen existing ones.

An increase in heavy freight vehicle traffic is liable to cause fatigue and the bridge must be safeguarded against this risk.

The need to improve traffic conditions may require alterations in bridge geometry (improving road alignment with possible repercussions for bridges).

I.5 SPECIAL FEATURES OF THE FIELD UNDER STUDY

I.5.1 A new and pressing problem

It is only in recent years that the scale of the problem of rehabilitation has been fully appreciated at the international level. Since about five years there have been some international initiatives, to study different aspects of the problems of rehabilitating and strengthening concrete structures. A number of countries have embarked on a general policy of road bridge rehabilitation, as it is described later in the report.

It may be useful to examine what has triggered off this international awareness.

The problem is not in fact a new one and there has already been much research and development in this area. What is new is the current relevance universally accorded to the range of tasks which go to make up bridge management in all its technical aspects.

This relevance derives from three factors:

- The first is the growing need to take account of all the economic aspects of the problems; this is discussed later on in this report.
- The second is the imperative, increasingly emphasized, of ensuring virtually failproof safety: this requirement has been reviewed at some length in the earlier OECD reports, notably in the study on bridge maintenance.
- The third, very different in nature, is associated with changing investment patterns (4). This is discussed hereunder.

Looking back over the last few centuries we find that short and highly productive periods with major capital investments tend to alternate with longer periods in which the main effort is to preserve the fruit of the previous one. In the case of highway networks, most industrialised countries had by the early 1980s completed their major motorway infrastructures and the related modernisation of their pre-existing highway networks. Levels of investment, while they are unlikely to fall to zero, are now cut back and the resources allocated over several decades to construction must in future be devoted to the upkeep of the existing infrastructure in their entirety, whether improved or newly constructed.

As is invariably the case during a period of intense building activities, highway authorities, engineers and experts were forced to leave aside other tasks, notably maintenance of pre-existing infrastructures, in order to cope with the investment effort. This is particularly true for bridges: as a result all too many old bridges, largely neglected, have deteriorated badly and either do not meet new safety standards or cannot cope with present traffic requirements.

Construction of new bridges on a vast scale over only a few decades necessitated recourse to new techniques, without all the implications involved in their use necessarily being realised at the outset. The longevity or perpetuity of bridges constructed using new and relatively untried techniques or materials cannot always be guaranteed. As a result many of the bridges built in the early years of the investment boom, are less reliable than their predecessors and have been left neglected for some considerable time; they are now in urgent need of repair.

Lastly, this construction boom substantially increased the number of bridges in service which has risen by several tens of thousands in Western Europe alone. To cope with this situation and to deal with the large-scale problems likely to result from only a few failures, it is necessary to institute a comprehensive policy that encompasses all aspects of bridge management. This implies that adequate resources be earmarked for the purpose. By and large all the resources liberated through the fall of investment should now be allocated to rehabilitation and repair.

In view of the current relevance of the problem, as highlighted above, a new market could emerge with considerable potential demand.

I.5.2 Scale of the problem

The quantitative information contained in Chapter II shows that there are about one million bridges (half of them in the United States) in OECD Member countries. Although the age of these bridges could not be ascertained for all the countries, it is likely that at least half of them have been built since 1950.

I.5.3 Difficulty of the economic approach

The institution of rehabilitation and strengthening policies implies some difficult economic choices. The difficulties of principle set out in the OECD report "Bridge Maintenance" (3) are compounded by the fact that the cost of repairing or strengthening a bridge must generally be worked out on a case-to-case basis; it is a function of the nature and the extent of damage, if any, the bridge's location, the volume of traffic, the feasibility or impracticality of works without obstructing traffic flow, etc. Depending on these factors the total cost can be up to ten times the pure construction costs.

It is hence well nigh impossible to estimate a priori the overall cost of rehabilitating or strengthening bridges along a route.

I.5.4. Technical difficulties

Repair and strengthening call for highly specialised techniques which have generally been developed very recently and whose possible effects have not yet been fully gauged. Moreover, it is necessary, before applying certain measures, to ascertain as fully as possible the present conditions of the structure. It is hence necessary to have available trained staff fully conversant with diagnostic and investigatory techniques as well as with the processes and techniques for carrying out the work. A high degree of expertise is called for. In case of errors in the choice of techniques or their inadequate application it is far more difficult to remedy the consequences than to repair the initial bridge defect.

I.5.5 Need to consider the global feature of the problem

It has already been stated in Section I.1 that rehabilitation and strengthening are only part of a broad panoply of measures, starting with inspection of the bridge, from the moment it is opened to traffic, and ending with its replacement.

During the normal life cycle of a bridge frequent maintenance operations are necessary. Rehabilitation works that are more costly are less frequent. These two types of activity are complementary and ensure long life spans. Furthermore, they are interactive insofar as inadequate maintenance accelerates the frequency of rehabilitation work.

Clearly, these two operations can only be implemented adequately if based on sound inspection data. Clearly, too, the requirement to strengthen a bridge or adapt it to new needs depends on the initial design, mainly as regards the bearing capacity, as well as forecasted future requirements. Needless to say that bridge design also has a direct impact on the facility of future repair or strengthening.

I.6 STRUCTURE OF REPORT

The report comprises the following chapters:

Chapter II: Rehabilitation and strengthening policies: present situation

This chapter has been drafted on the basis of information submitted by the countries represented on the working group. After briefly reviewing the extent of the problem in the different countries, which varies e.g. with the number of bridges in service, it goes on to describe present policies with particular reference to the inspection policies introduced after publication of the OECD report "Bridge Inspection" (1). It sets out how allowance is currently made for economic criteria. It then describes some rehabilitation and bridge improvement schemes initiated, holding up as an example the comprehensive bridge replacement programme launched in the United States. Lastly it mentions the special case of exceptional convoys and its implications for strengthening policy.

Chapter III: Repair and strengthening techniques: present situation

This chapter starts out by reviewing diagnostic and assessment techniques, examining in particular progress made since the publication of the OECD report "Bridge Inspection" (1) in which these techniques are inventoried. It goes on to describe the techniques habitually used for the repair of foundations, masonry, concrete and steel bridges, as well as of components and accessories. It details special techniques which did not figure in the report on "Bridge Maintenance" (3), referring readers to it in respect of the other techniques.

Chapter IV: Rehabilitation and strengthening policies: proposals

This chapter starts by reviewing the conditions which are likely in the foreseeable future to apply to the utilisation of bridges and which will influence appraisal of their service levels. Classifications are proposed which could serve to modulate the relative weights attached to the repair or replacement of bridge components on the basis, on the one hand, of the role of these components in the bridges and, on the other, of the bridges' location and function. Lastly it studies the impact of all the conditions and criteria set out above on decision-making and makes proposals as to how these factors can be taken into account. An annex to the report presents the results of an enquiry on the subjects covered in this chapter.

Chapter V: Recommendations for research

This chapter identifies areas where a further research effort would seem warranted, both as regards inspection and diagnostic techniques and repair and strengthening techniques.

Chapter VI: Conclusions and recommendations

This chapter summarises the report's main conclusions; it outlines the proposals and recommendations which the Working Group believes should be put forward by the OECD to national authorities at the policy and technical levels in the field of bridge management.

REFERENCES

1. OECD, ROAD RESEARCH. Bridge inspection. OECD, Paris, 1976.
2. OECD, ROAD RESEARCH, Evaluation of load-carrying capacity of bridges. OECD, Paris, 1979.
3. OECD, ROAD RESEARCH. Bridge maintenance. OECD, Paris, 1981.
4. A.B.E.M. (Association belge pour l'étude, l'essai et l'emploi des matériaux). Symposium "Gestion des Ponts". Brussels, October 1979.

Chapter II

REHABILITATION AND STRENGTHENING POLICIES: PRESENT SITUATION

II.1 SCOPE OF THE SITUATION

II.1.1 Background

For 25 years from 1950 to 1975 OECD countries built up a road and motorway network capable of meeting the challenge prompted by the enormous increase in motor vehicle traffic, and at the same time making good the time lost during the war years. Deviations and bypasses were constructed to replace existing roads. On newly constructed roads and motorways, grade separated intersections became the norm. Resources were mobilised on a massive scale in an all-out effort to build new bridges. During these years, however, maintenance of existing bridges was confined solely to urgent repairs.

This period of major bridge construction activities came to an end some time during the late 1970s and it became possible to take a closer look at the problems of maintaining the old bridge stock. In many cases the re-appraisal came none too early. However, the disquieting situation was fortunately tempered; new expertise and knowledge had become available and resources no longer required for the completion of the major road investment programme could be funnelled to this new task.

Throughout the OECD countries, a similar development occurred, varying in regard to its starting date and scale, depending in turn on actual conditions of the existing network. In the meantime, however, new issues emerged:

- the need to improve the existing network at the lowest overall cost;
- the need to cope with the increasing number of exceptional convoys.

The next two decades are likely to be characterised by the overriding requirement to preserve and efficiently manage the large stock built over the 25-year period quoted above. At the same time it will be necessary to rehabilitate and strengthen the older bridges which have been operated with insufficient maintenance and under excess loads for some 40 years.

As to the definitions of the terms "rehabilitation" and "strengthening" reference is made to Chapter I.

In a report of this nature, aimed at orientating future research on bridge management, it is useful to provide an overview of the size of the problem and to know the approximate total number of bridges in service in 1982 in participating countries. On the basis of the information received from these countries, it is possible to establish the following Table.

Table II.1

APPROXIMATE NUMBER OF BRIDGES

Country	Total Number	Minimum span	Comments
AUSTRALIA	30 000	N.K.	Source: PIARC
BELGIUM	5 300	5 m	- on the State road network
DENMARK	10 000	2 m	- of which 1900 on the State network; two thirds built since 1960
FINLAND	11 000	2 m	- on national roads and motorways; local community bridges excluded
FRANCE	110 000	2 m	- of which 55,000 with a span of more than 5 m: 13,000 on national network; 42,000 on local network
GERMANY	64 000	2 m	- of which 28,000 on federal network; 36,000 on local network
ITALY	30 000	N.K.	- on national network; local network is excluded
JAPAN	36 223	N.K.	- 12,571 on national network; 23,652 on local network, excluding community network
NETHERLANDS	4 000	5 m	- on national network; ing local network
NORWAY	19 400	2.5 m	- 7,800 on national network 8,600 on regional network 3,000 on local network
SPAIN	30 000	3 m	
SWEDEN	12 000	3 m	- 70 per cent built after the last world war
SWITZERLAND	8 500	5 m	- of which 2,735 (end of 1982) on the federal network and on major routes
TURKEY	4 682	10 m	- of which 3,086 on State roads and 1,596 on Provincial roads. (Structures of less than 10 m. span between side abutments are not considered highway bridges; they are categorised as drainage structures - village roads and settlement roads are excluded)
UNITED KINGDOM	155 000	0.90 m	- 13,000 on motorways and trunk roads - 142,000 on local networks
UNITED STATES	574 000	6.10 m	- of which 260,000 federal 314,000 local, half of which having been constructed since 1950

Even if the bridges under local jurisdiction which are not included in some national surveys are disregarded the predominance of the United States bridge stock in regard to the total number of bridges worldwide is obvious, especially considering the fact that bridges of a span below 6.10 m are not taken into account.

II.1.2 Inspection procedures

By and large all the countries have introduced bridge inspection systems along the lines advocated in the 1976 OECD report on bridge inspection or are in the course of so doing. In several countries computer assisted bridge management schemes are being developed. On the basis of national reports submitted to the Group the state of progress in the different countries can be summarised as follows.

Belgium. The guidelines are called "Règlement concernant la gestion des ouvrages d'art" (Directives concerning bridge management). They were published by the Bridges Bureau of the Ministry of Public Works, providing for several levels of inspection:

- routine inspection;
- annual surveying;
- "type A" general inspection at biannual intervals involving the preparation of an inspection report;
- "type B" general inspection, i.e. a specialised control carried out when the "type A" inspection reveals the need for it.

Given the degree of specialisation of the Belgian bridge management staff, the inspection directive does not specify the staff's needed expertise.

Denmark. Inspection is of three types:

- routine inspection which is effected on a daily or weekly basis by locally employed staff (workers or engineers). They check for visible damage and the control is made concurrently with general road inspection;
- a general inspection is carried out at least once every six years by a specialised engineer who schedules the date of the next inspection on the basis of his findings. It is carried out according to Road Instr. No. 8-20-02 which specifies in detail the procedure to be followed;
- indepth inspection in case of need.

Finland. The system can be summarised as follows:

- routine inspection is carried out by local staff, simultaneously with road inspection. Special attention is devoted to the bridges designed for loads that are lower than the actual traffic loads. This latter work is done by local district engineers;
- an annual control is performed by local technical staff and an inspection report is prepared;
- a general inspection is carried out every five years by specialised engineers of the district;
- a special inspection can be undertaken by the Technical Research Centre on behalf of the Road and Waterways Board.

France. There have been some major changes since the publication of the 1976 OECD report on bridge inspection. Inspection and repair of highway bridges on the national network come under a Technical Instruction approved by the Director of Highways on 19th December, 1979. There are no regulations covering inspection and repair of local bridges though the local authorities may apply the national rules and in fact generally do so.

The Technical Instruction is split into two parts:

- the first applies to all types of bridges, retaining walls, tunnels or even very high embankments. It lays down the broad lines of the inspection procedure to be followed;
- the second is made up of a number of separate sections, each devoted to a specific problem (foundations, structural elements, components, etc.).

The various operations stipulated are:

1. Daily surveillance
2. Yearly periodic inspection
3. Detailed five-year inspections
4. In-depth inspection in case of need

Germany. The information contained in the OECD report on bridge inspection is still valid. The inspection standards are those described in DIN1076. Each Land is responsible for ensuring that its bridges are kept in a good state of repair.

Italy. The 1976 OECD report states that bridge inspection is carried out according to Norms 6736-61 and 6737-61.

The Italian contribution to the Group confirms these procedures. It also indicates that there are three levels of inspection:

- daily controls by road surveillance staff;
- periodic inspection by specialised engineers;
- intensified inspection in case of need.

Japan. The organisation of bridge inspection is set out in some detail in the OECD report of 1976. The Japanese contribution to the Group did however point out an error in the OECD report as regards the number of bridges in Japan. Otherwise the inspection procedure as described in that report is correct.

Netherlands. As indicated in the 1976 OECD report, types and frequencies of inspections are as follows:

- coarse inspection: steel and pre-stressed bridges: every year
  reinforced concrete bridges: every year;
  foundations: every second year;
- detailed inspection: every tenth year;
- inspection of mobile bridges in accordance with special procedures.

Norway. The scheme described on pages 100-102 of the OECD report is in force; some adjustments will be introduced shortly, for example, detailed instructions for underwater inspections.

Spain. There are no modifications with regard to the system described in the 1976 OECD report.

Sweden. The Swedish contribution does not indicate any changes vis-à-vis the information contained in the previous OECD report. It specifies that inspections have been triennial since the 1940s. It notes that recent repairs have revealed more serious deterioration than appeared from the inspection reports. The Swedish submission particularly stresses the poor state of waterproofing layers and the upper part of the deck. The report concludes that inspection levels should be improved.

Switzerland. The Swiss submission does also not indicate any changes. Under the Federal system the cantons are free to organise the inspection procedures as they wish.

Turkey. The inspection system can be summarised as follows:

- routine inspection is carried out at frequent intervals by the maintenance technicians of the provincial divisions (Directorates of Divisions);
- annual control is carried out by the technicians of the Directorate of Bridge Maintenance, involving preparation of inventory inspection charts and the annual maintenance programme;
- on the basis of the annual maintenance programme, a very special inspection can be undertaken by a specialised engineer particularly for bridges designed for lower bearing capacity than the actual traffic loads;
- the maintenance operation is carried out by the engineers and the technicians of provincial divisions, according to the annual maintenance programme and the maintenance project of the bridges, prepared by the Directorate of Bridge Maintenance.
- inspection and maintenance of the bridges less than 10 m span is carried out within the road maintenance programme by the technicians of the Road Maintenance Department.

United Kingdom. The United Kingdom has been concerned quite early with this problem since a study, "Bridgeguard", was commissioned in 1967 to identify bridges with insufficient bearing capacity. The report of the MERRISON INQUIRY to investigate the bearing capacity and stability of steel box girder bridges was published in 1973.

Today United Kingdom procedures are officially laid down in Technical Memorandum B.E. 4-77 "Inspection of Highway Structures" which supplements the procedures described in the OECD report of July 1976. This memorandum summarises in a few pages what good management implies for the responsible highway authority. A sample worksheet is attached so that each bridge may be identified. A checklist serves to ensure that the inspector covers all important items.

An updated standard on the "Assessment of Highway Bridges and Structures", together with an advice note, has recently been established and issued for comment.

United States. Since half the bridges of the participating countries are on United States territory, it is proper that special attention be paid to how this country is approaching the problem.

The OECD report of July 1976 described the general organisation and there has been no change since.

It is in the United States that the most far-reaching decisions were taken in 1968 following the collapse of the Point Pleasant Bridge over the Ohio. It is worth recalling them hereunder briefly.

The Congress in its 1968 Federal-Aid-Highway-Act decided to establish a National Bridge Inspection Standard to ensure the safety of all bridges on the Federal aid highway system. The Act requires that an up-to-date inventory be kept of all bridges forming part of the system. Under the 1970 Federal Highway Act the Congress stipulated that bridges should be classified according to bearing capacity and safety for public use so as to determine replacement priorities. On 27th April, 1971 the National Bridge Inspection Standard (N.B.I.S.) was issued to meet the Congress' mandate and as of late 1973 most States had drawn up an inventory of bridges on the Federal aid highway system. Under the 1978 Surface Transportation Assistance Act Congress decided to extend the inventory and inspection programme to all bridges on public thoroughfares; the inventory was to be completed by 31st December, 1980.

As a result of these decisions the national inventory now contains the requisite data for:

- the virtual entirety of the 260,000 bridges on the Federal aid system; and
- 98 per cent of the 314,000 bridges on other public highways.

The NBIS requires that all bridges on public highways be inspected and that the inventory be updated every two years.

All the inventories, inspections and assessments are carried out under the responsibility of the State governments. The inventories are forwarded once a year to the Federal Highway Administration (FHWA) for inclusion in the National Bridge Inventory (NBI). The NBI is updated by the FHWA and used for a variety of purposes.

Lastly, the National Bridge Inspection Standard is supplemented by inspection manuals which provide advice to inspectors and even means for assessing the residual value of bridges with a view to replacement or repair.

The data collected served to establish a correlation between the age of bridges and the proportion of substandard bridges. Inter alia the following emerged:

- **the more rapid deterioration of bridges which are not included in the Federal Aid System compared to those in the system (Figure II.1);**
- the effect of the general economic climate bridge construction (Figure II.2).

The general conclusions drawn by the United States authorities were as follows:

"- a rational national bridge programme depends upon an accurate and comprehensive bridge inventory;
- the Federal Highway Administration holds national bridge data requirements to a minimum, but State and local officials can and often do add data items for their own management and planning purposes;
- the benefits and savings in human life resulting from the bridge inspection programme far outweigh programme costs;
- other financial benefits are very large;
- no State is yet taking full advantage of the bridge data available."

Figure II.1

PERCENT DEFICIENT BRIDGES VERSUS DECADE BUILT IN THE UNITED STATES
1981 Data

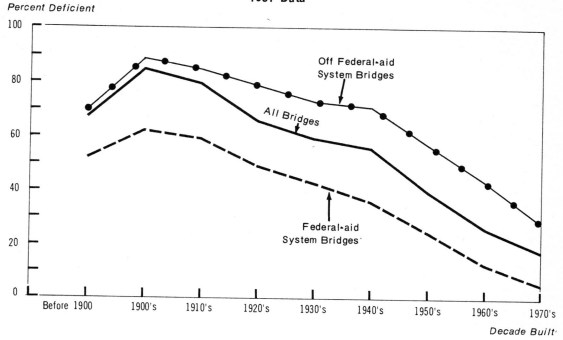

Figure II.2

NUMBER OF BRIDGES CONSTRUCTED BY DECADE IN THE UNITED STATES
1981 Data

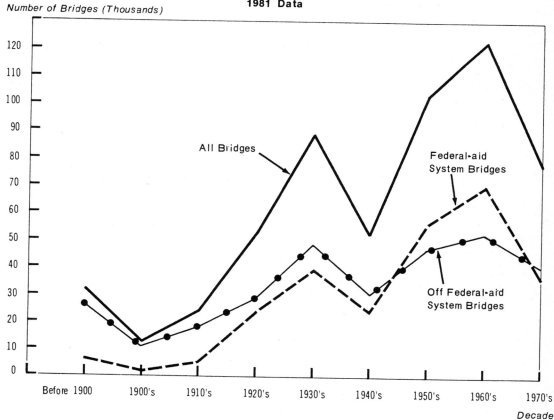

II.1.3 Overall review

There exist several levels of inspection:
- routine inspection performed by road surveillance staff in the course of their daily visits with reports submitted only in the case of obvious anomalies;
- detailed inspection by specialised technical staff at intervals of from two to six years according to country and type of bridge;
- intensified inspection in case of observed damage.

An important aspect concerns the costs for bridge inspection. Detailed information has been made available by Rhineland Palatinate, Germany. There are seven inspection teams for the annual control of a total bridge stock of 6,191 and a total bridge surface of about 2 million m^2(1979). Each team inspects about 130,000 m^2. It is composed of one engineer, one technician and one driver. They have a small delivery van carrying ladders and portable

Figure II.3

ROUTINE INSPECTION COSTS IN RHINELAND PALATINATE, GERMANY, IN 1979

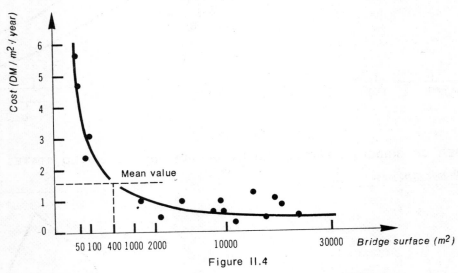

Figure II.4

COST FOR INSPECTION WITH MOBILE EQUIPMENT IN RHINELAND PALATINATE, GERMANY, IN 1979

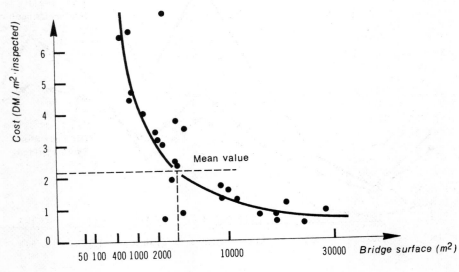

equipment (inflatable boat, tools, cover meter, rebound hammer, camera). The cost of such a team was DM213,000 per year in 1979, of which 80 per cent for labour costs. The average cost per square metre inspected was DM1.58 in 1979 but varied considerably with the surface of the bridge. Figure II.3 illustrates this variation and Figure II.4 presents the cost variation as a function of inspected bridge surface in the case of using mobile inspection equipment on an average every six years (mean value = DM2.19). For the 1,500,000 m^2 necessitating mobile bridge lifts the average total cost was therefore $1.58 + \frac{2.19}{6}$ = DM1.95 m^2/year.

The sum of the expenditure for the above two methods of inspection gives the total cost of inspection.

In conclusion, it can be said that although national inspection policies agree in many respects, they differ as to the way the inspection data are used; there tend to be two schools of thought:
- in the one case, all the data are centralised and stored in data banks, this is the case for example in Belgium and the United States;
- in the other, management is decentralised and data are stored at the local level of bridge management; this is the case for France and many other countries.

The submissions received do not reveal a general trend. It is likely, however, that with the development of computerised data processing, authorities can be kept informed about the state of their bridges on both local and national networks. This information can be beneficial to decisions about the allocation of resources.

II.2 FINANCIAL AND ECONOMIC ASPECTS

II.2.1 Direct costs borne by the bridge authority

The financial and economic aspects are important in determining policy for rehabilitating, strengthening and replacing bridges. The life of a bridge or of a particular component is a crucial factor. A bridge's lifespan depends on a number of factors:
- the initial design and its quality;
- the nature and quality of the materials;
- the care taken in construction;
- traffic loads carried, particularly heavy traffic;
- the quality of maintenance;
- the severity of the climate, the effects of water, etc.

The bridge authority has to manage the bridges as constructed and as subjected to road traffic. It can reduce permissible loads in case of danger. It can act by maintaining or strengthening bridge elements. In some cases it may intervene to alleviate the effects of the surroundings of the bridge.

Providing the inspections are carried out regularly and the records are properly updated, the necessary data should be available to decide on the most appropriate repairs. The actions to be taken will depend on:
- whether the annual appropriations are sufficient to carry out the work revealed by the inspection to be indispensable;
- whether the annual appropriation fall short or even far short of what is deemed necessary.

In the years when the bridge stock is being built up, the management authority generally finds itself in the first situation; such was the case until recently for much of the present stock was constructed after World War II. This situation has begun to change now that these bridges are ageing.

A Danish study indicates that needs grow as bridges get older. Taking average bridge life as 60 years, the Danish submission based on a survey of 2,000 bridges, gives the following maintenance expenditure:

Table II.2

ANTICIPATED REPLACEMENT AND
MAINTENANCE COSTS IN DENMARK

Year	Number of bridges	Total replacement cost (10^9 \$)	Average bridge age	Maintenance expenditure (10^6 \$)	Maintenance/replacement cost ratio
1980	1817	1.0	20	12.6	13 per 1000
1990	1970	1.2	28	24.2	20 per 1000
2000	2010	1.2	37	31.9	25 per 1000

These figures (in 1982 \$) include the cost of replacing bridges that are no longer serviceable, but exclude costs due to longer travel distances and road user delays.

It is thus essential to improve maintenance techniques so as to limit this rising expenditure which the budget authorities would probably refuse to countenance. This situation pertains throughout Europe and possibly even in Japan. It is different in the United States where a vast bridge rehabilitation and replacement programme has been launched; the United States situation is reviewed in II.1.2 and II.2.8.

II.2.2 Basic considerations when assessing future costs

Bridge management authorities need to be able to work in the present while preparing the future. It is essential to be in a position to forecast future maintenance costs and establish how they will be apportioned. To facilitate this task, it may be useful to break costs down as follows:

- the cost of maintenance operations. Here we refer to routine maintenance which will vary with the size of the bridge and is relatively independent of age;
- the cost of repairing though not replacing bridge elements: paint, treating and repairing pavement and sidewalks, repointing masonry;
- the cost of replacement. These costs may cover replacement of the bridge elements or even of the bridge itself.

Some bridge elements (surfacings, waterproofing, expansion joints, parapets, guard rails, and even bridge bearings) are replaced from time to time. The replacement of other elements (bridge decks, retaining walls, piers, abutments and foundations) is exceptional and replacement is generally effected in the course of reconstruction or bridge widening.

II.2.3 Lifespan of bridge elements

Studies carried out in various countries reveal that the bridge elements that are frequently repaired or replaced represent between 12 per cent (France) and 20 per cent (Denmark) of the initial outlay on bridge construction. A French study has shown that this 12 per cent can be broken down as follows:

- expansion joints   1%
- bearings   1%
- water proofing   2.3%
- parapets and guard rails 3%
- surfacings, drainage, etc. 4.7%

         total   12%

These percentages are based on a survey of newly constructed bridges and do not apply to bridges in need of repair. Replacing these elements in a bridge in service accounts for the virtual entirety of ordinary maintenance appropriations. Protective treatment and repairs to structural elements include crack repair, repairs to the face-walls, protection of foundations against water erosion. Particular mention should be made of the pre-stressing cables, where damage is generally the result of design or construction faults rather than of poor maintenance of correctly designed and constructed components. The Danish submission contains the following Table:

Table II.3

ASSUMED SERVICEABILITY EXPECTANCY OF
ROAD BRIDGE ELEMENTS IN DENMARK

Elements	Surfacing Water proofing Rails Bearings Joints etc.	Bridge decks Columns Foundations Retaining walls etc.
Cost	20% of the total replacement cost	80% of the total replacement cost
Expected average	20 years	60 years
Expected minimum	5 years	20 years
Expected maximum	35 years	120 years
Mean serviceability expectancy	20 years	64 years
Standard deviation	6 years	20 years

The above cost figures of 20 per cent and 80 per cent are based on studies of bridges in need of repair and take into account ancillary repair costs. On the other hand, costs for traffic diversions are not included.

The Danish Study was of course undertaken in a country where weather conditions are particularly harsh. Throughout the four - five month winter temperatures oscillate around 0°C and humidity is high; large quantities of de-icing salts are applied. In consequence the Danish submission predicts a very high outlay on bridge replacement as from 1990.

In the other countries the lifespan of structural elements is between 100 and 150 years (See also answers to an enquiry, summarised in the Annex).

A detailed French survey covering 236 bridges on major routes showed that:

- 43 bridges required major repairs (18 per cent);
- 5 bridges required reconstruction (2 per cent);
- 3 bridges required widening (1 per cent).

The costs involved for each of the three categories are estimated at (1982 $):

- $155 per sq.m for the 43 bridges requiring repair (i.e. 10 per cent reconstruction costs);
- $1,550 per sq.m for each of the five bridges requiring reconstruction;
- $900 per sq.m for each of the three bridges to be widened (i.e. 60 per cent of reconstruction costs).

This gives an average outlay of $285 per sq.m to take account of unforeseen eventualities so that for the bridge stock as a whole the total outlay per sq.m would be (18 + 2 + 1) x 285/100, i.e. $60 per sq.m. This outlay on repair or on maintaining service levels would be spread over some 20 years. The survey also indicates that the 236 bridges which were examined and which are on the national network are broadly speaking in good condition; otherwise the average outlay would be higher.

II.2.4 Bases for calculating replacement costs

The cost of replacing bridge elements should include ancillary (indirect) expenditure which is often greater than the replacement as such, e.g.:

- demolition;
- replacing parts which are sound but connected with damaged components;
- traffic diversions;
- economic consequences of such diversions (if these can be assessed).

Replacement costs should be calculated as of the date when a component has deteriorated to a point where replacement is essential.

The role of repairs is to defer replacement; it is almost always possible to repair, even though the cost may be high and this tends to be the practice where replacement would have inacceptable consequences for traffic. However, the relative costs of replacement and repair must be weighed up.

In the United Kingdom the TRRL(*) has carried out a study to assess traffic delays and associated road user costs (but excluding accident costs).

The main application of the QUADRO (Queues and Delays at Roadworks) model is the economic appraisal of delays during future maintenance work, given a knowledge of how the traffic is likely to be managed. The model can also be used to assess the effects of traffic delays occurring during road improvements.

QUADRO models a simple network consisting of a main route containing the roadworks site and a single alternative route by which the driver can divert around the works. This diversion route need not be an actual road

---

*) PARKINSON, MN and PHILBRICK, MJ. Assessing the total cost of major maintenance works. Proc. PTRC Annual Meeting, Summer. July 1979.

but can be a "notional" route representing a combination of all probable diversion routes for the site.

The delays calculated by the model are a result of queueing in the approaches to the site, increased journey times through the site and increased journey times on the diversion routes, both for the diverted traffic and for traffic which normally uses that route. The model also calculates additional delays and costs caused by vehicle breakdowns within the site.

The road user costs calculated in the model are divided into two parts: the delay costs arising solely from increased journey times, and the extra vehicle operating costs arising from both reduced speeds and increased distance travelled.

Vehicle breakdowns within the site are allowed for by modelling the possible behaviour of drivers encountering unexpected queues or queues which are longer than normal. A breakdown rate for the site is calculated, and the delays and associated costs on the main and diversion route arising from these breakdowns are evaluated.

II.2.5 Replacement costs of bridge elements

Some data are available on the cost of replacing a bridge element compared with the cost of a new component when the bridge is first constructed. In the following some values for the multiplying cost coefficients are given:

- changing an expansion joint costs two to three times the price of a new joint if automobile traffic can be diverted over half the pavement;
- guard rails, etc., are more easily changed; the coefficient falls to 1.5;
- for the waterproofing layer the coefficient is three because the pavement must also be taken up, even if it is in good condition. If account is taken of the fact that the pavement is generally in a poor state of repair when the waterproofing layer is changed, the cost coefficient for the waterproofing and the pavement combined falls to around 1.3;
- in changing a bearing account should be taken of the cost and implications of lifting it. The following coefficients have been noted:
  - about ten times the cost of a new bearing in favourable conditions;
  - 40 times the cost in the majority of cases;
  - hundreds of times the cost in exceptional cases.

As regards the structural elements it is impossible to give even approximate estimates. If the new bridge were reconstructed adjacent to the first one, the incremental cost could be negligeable. If, on the other hand, temporary structures have to be erected and traffic diverted, the cost could be enormous, giving a coefficient which may be as much as about ten.

All these considerations alter the relationships between the replacement values of the various elements which can be quite different from those pertaining in a bridge constructed on a new site.

Several national submissions point out that the basic problem of maintenance is one of durability. The only way of improving the durability of bridges already built is to inspect them carefully and repair them in due time.

II.2.6 Consequences of appropriation ceilings

Since the object of rehabilitation is to defer replacement it is necessary to compare between:

- the cost of a repair which enables "n" years to be gained on replacement and
- the cost of replacement.

The maintenance budget being fixed and unlikely to be increased, rehabilitation can allow two bridges to be brought back into service instead of one replacement. This question was treated in Chapter IV of the 1981 OECD report on bridge maintenance; there is little to add except to recall that the costs to be taken into account should make due allowance for the multiplying cost coefficients specified in II.2.5 above.

II.2.7 Desirable funding

In conclusion, on the basis of the information received, amounts to be earmarked for bridge maintenance and replacement represent on average:

- 0.3 per cent of the invested capital(*) for routine maintenance and minor repairs to avoid substantial outlays at a later date;
- 1.4 per cent of invested capital(*) to rehabilitate the bridges, strengthen them, replace those which are outworn and - lastly - meet the demands of exceptional convoys, an issue which is discussed in II.4;
- with a new bridge stock the ratio of 1.4 per cent can be somewhat reduced. It is not possible to suggest a definite lower limit since this depends on the type of bridge. With massive bridges in reinforced or pre-stressed concrete the ratio falls below 1 per cent. In the case of slender bridge structures, the ratio of 1.4 per cent is insufficient.

If no or insufficient funds are earmarked for bridge rehabilitation, it is likely that no negative effects will occur during the first year and sometimes even after several years. Nevertheless, by doing this the existing capital investment is losing value, and the medium-term consequences are always very severe. It should be noted that the failure of a bridge or, even only its closure to traffic without a complete bridge failure can have enormous consequences for the economy of a region if such an event occurs suddenly. The resulting expenditure would then be much greater than the economies made by neglecting necessary repairs.

II.2.8 The United States programme

Particular attention has been given to this problem in the United States where a vast bridge rehabilitation and replacement programme is underway.

---

*) Cost of constructing a bridge element in the case of a new structure, excluding costs for demolition, traffic diversions, etc.

Total needs were set against available resources. Needs are very considerable since to strengthen or replace 200,000 bridges requires $50 billion. Available resources are however substantial (in 1981 total funds available amounted to $2.5 thousand million) and the strengthening programme should, if all goes well, be completed some 20 years from now (Figure II.5).

Highest priority is given to the repair of 39.000 bridges whose wearing courses are badly damaged through the use of de-icing salts and some $9.6 thousand million have been provisionally earmarked for the purpose.

Another aspect of the programme is the strengthening of 127,000 bridges which are below standard. Some $1 million per bridge has been estimated for bridges in urban areas.

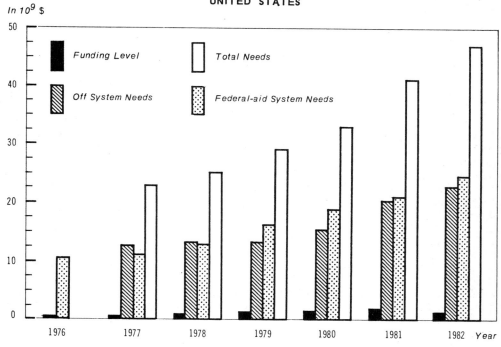

Figure II.5

COMPARISON BETWEEN FUNDING LEVELS AND NEEDS FOR THE HIGHWAY BRIDGE REPLACEMENT AND REHABILITATION PROGRAM UNITED STATES

II.3 NEEDS ASSOCIATED WITH NORMAL TRAFFIC

II.3.1 Adaptation of bridge structures

Adapting existing bridges to prevailing traffic requirements is a major task for road administrations.

The greater part of the needs are geometric since there has been a proportionately larger increase in the number of vehicles in circulation than in their weight.

As regards axle loads international harmonization is pursued. Lorries cross frontiers and this is obviously not without problem for old bridges. Happily our predecessors tended to be generous in their design standards as regards the load carrying bridge elements and it has been possible to take advantage of this safety margin. This is not, however, always the case and for some one-fifth of bridges in Europe and around one-third in the United States either strengthening or repair is required.

II.3.2 Strengthening

In the case of strengthening work, one should take into account notably:

- the load to be carried;
- traffic volume and composition including provisions for cycling traffic in some countries;
- the inconvenience caused by the strengthening works to other types of traffic on or under the bridge. This inconvenience often plays a prime role in determining which bridge strengthening or widening technique should be employed.
- the characteristics of the structure.

II.4 EFFECTS OF EXCEPTIONAL CONVOYS

Exceptional convoys are those that involve the transport of indivisable loads whose weight and/or size exceed the maximum authorised limits.

Several countries have introduced measures to cope with such transports in the national context. These regulations differ from one country to the other, depending on the geographic situation, the existence of a waterway network on the different distances between the production sites of such abnormal loads and the location of sea harbours.

The Group noted the general similarity of the national procedures in assessing the exceptional transport needs and of the solutions adopted. The information made available to the Group was however too fragmentary to allow for an indepth study and for drawing conclusions that would be valid for all countries.

In view of the great importance of this issue the Group wishes to recommend the initiation of a special Working Party to study this problem in depth.

Chapter III
REPAIR AND STRENGTHENING TECHNIQUES: PRESENT SITUATION

III.1 INTRODUCTION

The Chapter covers only the more important techniques used today for bridge repair and strengthening. The criteria for selecting techniques are as follows:

- their importance for the preservation and load carrying capacity of the structure;
- their relative novelty in comparison to traditional, well known, techniques.

Detailed information on and description of well-known bridge maintenance techniques which were included in the 1981 OECD Report on Bridge Maintenance(1)* are not repeated here.

This Chapter comprises 9 sub-chapters:

- Condition Evaluation
- Repair of Foundations
- Masonry Structures
- Concrete Structures
- Steel Structures
- Composite Structures and Moveable Bridges
- Components, Accessories
- Seismic Retrofitting
- Structural Analysis and Load Testing.

A brief description of the nine major areas covered in the Chapter is as follows:

Condition Evaluation (Structures and Materials)

Special investigations or studies carried out by Member countries and by CEB (European Committee for Concrete) are listed, assessment techniques and special tests are discussed and their merits and limits are given in tabular form. A guide for concrete defects is given and underwater inspection described in more detail.

Repair of Foundations

Methods for repairing and strengthening of foundations are described

---

* See list of references at the end of this Chapter.

emphasizing underwater bridge structures and examples of repair work as used in Member countries are given.

Masonry Structures

Rehabilitation and strengthening methods, especially those for masonry arch bridges, are mentioned and the importance of foundations is underlined.

Concrete Structures

The scope of the problem is outlined, repair and strengthening strategy discussed from the technical point of view and the following subjects are treated: chloride contaminated concrete; cathodic protection of concrete decks; replacement of corroded reinforcing steel; external reinforcement and supplementary prestressing. Two examples of strengthening operations are described.

Steel Structures

Guidance is given for typical defects in steel structures. Rehabilitation and strengthening methods for iron and steel bridges and repair methods for cracks in steel members are presented. Guidance for deck replacement in older steel bridges is given. A strengthening method of old truss bridges by using a steel arch superposition scheme is introduced and possible problems related to suspension and cable stayed bridges are discussed.

Composite Structures and Moveable Bridges

Problems with this type of bridges are generally similar to those described in foregoing sub-chapters; special problems are briefly mentioned.

Components and Accessories

The operational life of such elements is usually shorter than that of the main structural elements. Replacement is the usual method of rehabilitation.

Seismic Retrofitting

The following aspects are discussed and some are treated in more detail: bearings and expansion joints, columns, piers and footings, abutments and liquefaction of foundation soil.

Structural Analysis and Load Testing

The point is stressed that advanced methods of analysis and load testing may show that some bridges have a higher load carrying capacity than anticipated in the original design calculations.

III.2 CONDITION EVALUATION (STRUCTURES AND MATERIALS)

The evaluation of a bridge which has been in use for some years must be based on the assessment of the <u>actual</u> condition of the structure. Evaluation is dependent upon systematic and detailed bridge inspection and the expert assessment of data. Full information is required on the design of the bridge, on the properties of the materials used and on any deterioration, defect or damage. The original design concept must be understood and the bridge should be examined to determine whether it is behaving as expected.

III.2.1 Results of inspection

Systematic bridge inspection is being practiced in most Member countries. Condition evaluation and repair needs are usually determined from bridge inspection reports. The reports are essentially an assessment of the condition of the structure and do not usually refer to its load carrying capacity.

Special inspections may be made:

- when an abnormally heavy load is moved and it is necessary to check for load carrying capacity
- when a major principal inspection indicates that further investigation is needed
- after special events such as: heavy storms, floods, earthquakes, fire, etc.

From time to time, special investigations or studies have been undertaken (see Chapter II.1.2). Noteworthy are an ongoing study investigating cracks in prestressed concrete bridges in Germany and the special load testing programme to define the actual load carrying capacity of older bridges in Ontario (Canada).

III.2.2 Assessment and special tests

Previous OECD Reports (2, 3) have treated in detail the subject of assessment techniques. In order to avoid repetition, only new techniques or those that have been developed further are presented here.

Non-destructive testing can make a valuable contribution to the investigation of many problems that occur in bridges. However, the nature of its contribution should be properly understood. More commonly it will provide the engineer with some additional evidence about a problem that he already suspects. Sometimes it may do no more than confirm his suspicion. It should not therefore be regarded as a diagnostic technique in itself but as a part of a broad approach to the investigation of a problem. The drawings, the records of construction and testing, inspection records and visual examination sometimes augmented by strain and deflection measurements and non-destructive testing should be regarded as complementary to each other. Hypotheses about the mode of behaviour or the type of fault can then be tested against the assembled evidence. Non-destructive testing is probably most valuable when used in a supporting role.

A strict definition of non-destructive testing would confine it to techniques which do not in any way change the properties of the structure or alter it physically. These techniques would include the following as applicable for different materials:

- measurement of the velocity or attenuation of ultrasonic pulses transmitted through the structure (or its components);
- the use of ultrasonic pulses to obtain reflections from internal flaws;
- the response to magnetic or ultra-magnetic fields;
- the detection of accoustic signals due to crack extension when the structure is loaded;
- measurement of electrical properties;
- surface penetration methods;
- the use of radiography to examine the structure in depth.

Such a definition of non-destructive testing is somewhat restrictive when applied to bridge assessment and the term non-destructive is widened in this chapter to include some tests involving removal of small samples by drilling, cutting or scraping but always on a minor scale which will not impair the integrity of the structure or its components.

The engineer's approval should always be obtained before any samples are removed. He will have to consider the likely value of the results in relation to possible damage to the structure and whether indirect methods of assessing might not be more appropriate.

The tests to be used depend very much upon the circumstances and condition of the particular bridge and upon the kind of problem being investigated. For instance, if corrosion of reinforcement in concrete bridges is suspected then measurement of electrical potential would be supplemented by measurement of the concrete cover, and permeability and chloride content of the concrete.

The use of radiography or radiometry will need the employment of specialists. The other tests could be carried out by a highway authority but would depend upon its having suitable staff and equipment including laboratory facilities.

The application of many of the non-destructive techniques for the examination of faults in bridges is comparatively new. Until there is experience of the application of those techniques to bridges on a wider scale it is difficult to make firm recommendations and it will be necessary for the engineer to consider each particular case on its merits.

Available assessment techniques for reinforced and prestressed concrete structures are listed in Tables III.1 and III.2. The merits and limits of the different techniques listed in Table III.2 are considered below:

Visual Examination

Visual evidence of distress may be given by cracking of the concrete, by opening of construction joints, by spalling, by damage due to tendon fracture (i.e. through release of strain energy), by deformation or by rust staining. Wire fractures may be detected by careful examination of the sleeving around the strands in externally post-tensioned bridges. The presence of voids and corrosion of tendons in the ducts of post-tensioned structures will normally only in exceptional circumstances be detected by visual inspection. In general, visual distress will often not become apparent until deterioration has reached a fairly advanced stage.

TABLE III.1

CLASSIFICATION OF TECHNIQUES FOR EXAMINING REINFORCED CONCRETE STRUCTURES FOR CORROSION

Technique	Direct	Indirect	Non-destructive	Semi-destructive	Destructive	Corrosion Rate	Corrosion Defect	Corrosion Causes	Existing structures	New structures
Visual investigation	•						•		•	
Weight loss	•					•				
Pit depth	•				•	•			•	
Electrical resistance probe	•		•			•				•
Linear polarisation	•		•		•	•				
Half cell potential	•		•				•			•
Carbonation		•	•		•			•	•	•
Covermeter		•	•					•	•	
Chloride analysis		•		•				•	•	
Cement content		•		•				•	•	
Resistivity		•		•				•	•	
Moisture content		•						•	•	
Water absorption		•			•			•	•	
Concrete strength		•			•			•	•	
" permeability		•			•			•	•	
Delamination		•	•				•		•	
Ultra-sonic methods		•	•						•	
Impact hammer		•	•					•	•	
Gamma radiography		•	•					•	•	
X-ray photography		•	•				•	•	•	
Windsor probe		•		•				•	•	
Coring		•			•			•	•	

TABLE III.2

METHODS FOR THE EXAMINATION OR MONITORING OF POST-TENSIONED CONCRETE BRIDGES
(x-possibilities under consideration or development)

Kind of examination	Method		Information obtained
INSPECTION	Visual examination of structure		Cracking of concrete; tendon fractures (when there is visual evidence)
NON-DESTRUCTIVE TESTING (Examination through the surrounding concrete)	Radiography		Voids, tendon fractures
	Radiometry	x	Voids
	Ultrasonic pulses		Voids
	Magnetic field	x	Tendon failures
TESTING (Direct examination or measurement)	Access holes drilled into ducts for:		
	- endoscopic examination		Visual examination of condition
	- pressure/volume measurements		occurence and size of voids Air leakage
	- sampling		Carbonation, pH-value and chloride content
	Cutting and drilling to produce local relaxation	x	Residual stress
MONITORING OF CHANGE OF BEHAVIOUR	Change of strain	x	
	Change of longitudinal profile	x	Change in residual stresses
	Redistribution of support Reactions	x	
	Accoustic emission	x	Tendon fractures
	Vibrational characteristics	x	Cracking, debonding between components, wire fractures
	Proof loading	x	Loss of stress (in some circumstances only)

38

Non-destructive Testing

At present no satisfactory non-destructive testing technique is available which will detect voids mainly in ducts and corrosion and fracture of the prestressing tendons without damaging the structure. Radiography can be used for detecting voids and, if circumstances are ideal, for severe corrosion and tendon fractures. Normally its use is restricted by the geometry of the structure and/or its components, cost and often the hazards of using ionising radiation. Radiometric scanning may overcome some of the disadvantages of radiography (in principle such a technique could give a three dimensional picture), but its use would often still be restricted by the geometry of the structure and cost. Little success has been shown so far in attempts to use ultrasonic techniques for detecting voids in ducts. A method for detecting wire fractures, based on the disturbance generated by a fractured wire to a magnetic field, is in the development stage in the United States.

Testing by Direct Examination and Measurement

For post-tensioned structures, examination and sampling of conditions inside the duct, by drilling holes, provides at present the most positive method of obtaining information about voids and the risk of corrosion. Typically, these holes will be drilled at those locations that can be normally considered to constitute dangerous zones such as anchorage zones, high and sometimes low points of the cables. The information gathered relates however only to the parts of the duct examined and to the tendons or wires that are exposed. Ideally a sampling technique should allow the results to be applied with confidence to a whole structure or at least to its most important components. At present, the hole drilling technique should be regarded as a viable method for investigating structures where problems are known to exist.

The only likely direct method of determining the magnitude of the permanent stress in the concrete is by local relaxation of stress. Where there is already a crack or construction joint which opens under load, the stress may be calculated. Otherwise it will be necessary to produce relaxation by cutting, drilling or by removal of samples.

Development of Monitoring Systems

Monitoring may consist of periodic measurements to detect any change in the behaviour of the structure or continuous surveillance to detect the occurrence of fractures. Tendon fracture or loss of prestress would be revealed by changes in the following parameters:

- strain in the concrete
- deformation of the component
- distribution of the reactions at supports
- vibration response characteristics
- accoustic noise within the component (fractures only).

The aim should be to detect any trouble at a significantly earlier stage than would be possible by visual inspection. This requires that the information needed can be distinguished from the effects such as of temperature changes, creep, settlement, etc.

Since the cost of using the techniques described above is high, the choice of structure to be examined in detail should be based on the role of corrosion and the response of the structure to loss of prestress. A preliminary task therefore is to obtain as much design and structural information about the structure as possible.

III.2.3 Guide for concrete defects(4)

Deterioration of Concrete

- local disorders
  - honey combing, segregation
  - spalling, scaling, abrasion
- leaching
  - efflorences, stalactites
  - water seepage
- chemical or physical attack, possibly due to the quality of the cement and aggregates used in the concrete

Deterioration of Steel

- exposed reinforcement in reinforced or prestressed concrete
  - corrosion
  - rupture
- anchorages or covers
- faulty grouting, accumulation of water in ducts
  - corrosion
  - wires released, broken
  - tendons released, broken
- rupture of sockets, bars (e.g. stress corrosion)

Abnormal Deformations or Movements

- abnormal vertical deflection
- abnormal horizontal deflection
- abnormal vibration
  - due to traffic
       to wind
- abnormal movement at supports
  - at fixed ends
  - or blocking
- abnormal deformation which makes the structure unsuitable for its designed use

Cracks

- location and description
  - small cracks less than 0.1 mm
  - medium cracks from 0.1 to 0.3 mm
  - large cracks more than 0.3 mm
  - active cracks due to live loads or temperature
  - cracks changing the structural system

Closure, Partial or Total Collapse

- closure by a decision anticipating the risk of aggravation of defects

- accident by the phenomena of instability
    - buckling
    - tilting
- rupture with partial collapse
- rupture with total collapse.

III.2.4 Underwater inspection(5)

The quality of inspection under water should be equal to the quality of inspection above water. Underwater inspection differs from inspection above water in that more hazards are involved; the inspection mobility, visibility and physical endurances are affected and obtaining samples is more difficult. Inspections must be conducted by experienced, competent divers. Scuba and surface supplied air diving are the two methods of diving operations most applicable to bridge inspection work.

The planning for underwater inspection should include consideration of the season, flow, depth, turbidity, visibility, materials, and type and condition of the structure.

Cleaning aquatic growth from the underwater portions of bridge substructures is usually required to facilitate inspections. The amount of cleaning necessary depends on the amount of growth and the type of inspection being made. Cleaning of surfaces under water can be accomplished with divers knives, hand tools, scrapers, power chippers, grinders and brushes, and sandblasting and water jets. Highpressure water blasters must be used carefully to avoid damage to the structure and injury to the operators and others.

Underwater damage or deterioration can be identified by several methods of detection, including:

- visual inspection as the primary method
- tactile examination
- settlement or loss of alignment of the structure

In some cases, more sophisticated instruments are required for the detection of underwater flaws, damage or deterioration; these include echo sounders, ultrasonic thickness gauges, computerised tomography, and closed-circuit underwater television.

The most obvious limitation to visual inspection is water clarity. When turbidity is high, underwater visibility can be zero. Visibility may be improved if a clear water mask is attached to the face plate of diving gear. Small plastic bags (0.01 m3) can be used as a clear water lens to inspect H piles and other irregular shapes.

Echo sounders, specifically fathometers, are effective in checking scour in the stream bed adjacent to a bridge. Undermining of piers or abutments cannot be adequately detected with an echo sounder; when undermining or undercutting is suspected, there is no substitute for visual inspection.

However, since in many cases scouring is greatest during high water flows when it is too dangerous for divers to do the inspecting, the actual damage to foundations done by scouring cannot easily be determined.

Computerised tomography is a recent addition to underwater equipment. With this method, voids can be located, steel location can be pinpointed, and sections can be indicated accurately.

Figure III.1'
SCANNING PROCEDURE FOR COMPUTERIZED TOMOGRAPHY

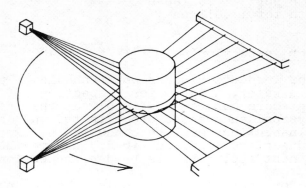

A gamma ray source is collimated to form a flat fan of rays that are attenuated as they pass through the object to a set of detectors. The source detector apparatus is rotated to give a series of such projections, all through the same cross-section.

Documentation of observations on video tape is standard practice in some underwater inspections. Cameras are designed to be handheld or mounted on the divers headgear. As a documentation tool, television offers no particular advantage over still or cine photography. It is common practice to supplement television with still photographs. Television does offer the advantage of real-time display to the surface and real-time quality control of the video image.

III.3 REPAIR OF FOUNDATIONS

III.3.1 Repair methods

A general method for repair and/or strengthening of foundations does not exist. Different cases have to be analysed individually and require special investigations. Most repair work is in the category of protection and strengthening. Some examples are given below:

- scour and erosion protection,
- repair of foundations built on soft ground subjected to erosion,
- strengthening of foundations affected by extra load on existing piles due to settlement of soft ground under the foundation,
- remedying the effects of horizontal movements of abutments built on soft ground,
- strengthening of foundations due to widening of a channel or a road,
- extending existing foundations,
- repair of underwater bridge structures.

The combinations of conditions requiring repairs and the causes of deficiencies of underwater bridge structures are too numerous to be able to specify exactly each repair case. Given the wide range of materials and repair techniques, the choice of the most appropriate techniques is not easy. Table III.3 lists, according to the nature of the problem and

TABLE III.3

REPAIRS AND PREVENTIVE MEASURES

Type of repairs. (underwater and in splash zone)	Scour	Deterioration			Damage (structural)			Structural failure			Foundation distress		
		Con-crete	Steel	Timber	Con-crete	Steel	Timber	Con-crete	Steel	Timber	Con-crete	Steel	Timber
Replacement of material	•												
Sheet piling	•												
Training works	•												
Modification of the structure	•											•	•
Epoxy injection		•			•			•	•	•	•		
Quick setting cement		•			•								
Epoxy mortar		•			•								
Placing concrete underwater — Underwater bucket		•			•			•			•		
Placing concrete underwater — Tremie concrete		•	•	•	•	•	•	•	•	•	•		
Placing concrete underwater — Pumped concrete		•			•			•			•		
Placing concrete underwater — Prepacked or bagged concrete		•			•			•			•		
Protective coatings		•	•			•							
Cathodic protection			•			•							
Splicing new steel sections			•			•			•				
Pile jackets		•	•		•	•	•	•	•	•			
Wood treatment				•									
Flexible and rigid barriers		•	•	•				•					

43

the type of repair, some of the techniques and preventive measures currently performed in the United States under water and in the splash zone.

Underwater Work

Methods used by divers to perform underwater (down to about 18 m) sealing and repairing of cracks by epoxy injections are similar to methods used above water, except that the epoxy surface sealer takes several days to harden sufficiently to withstand injection pressure. Epoxies for underwater use must be water-insensitive. Before the application of epoxy surface sealer, cleansing is necessary. If oil or other contaminants are present in the cracks, and the epoxy is used for restoring the strength of the cracked concrete pier or pile instead of simply blocking the free entry of water in the crack, bonding will be improved by mixing detergents or special chemicals with the blast water to clean the crack interiors. After all cracks are prepared and sealed and the nipples positioned, the low-viscosity epoxy adhesive is injected under pressure into the crack network. A surface-mounted, positive-displacement pump is used to dispense the adhesive's two components to the submerged injection sites where the adhesive is mixed in the injection head as it is pressure-pumped into concrete. Water temperatures must be above 4°C. The adhesive cures to full strength in about 7 days. Cracks up to 6 mm wide may be sealed with pure (unfilled) epoxy resin. For wider cracks, the addition of a filler is generally required.

Various methods are currently used to prevent the corrosion of steel piling in seawater, including application of protective coatings, cathodic protection, encasement of the steel in concrete, or a combination of these procedures.

Cathodic Protection

This is an electrochemical method that is effective in preventing corrosion of new or existing steel piling located in seawater, freshwater, or soil by making the metal surface cathodic with respect to its environment. There are two basic methods of applying cathodic protection:

a) The _galvanic anode system_ uses anodes, usually made of magnesium, aluminium or zinc. Because of difference in potential between the anode and the pile, a battery (corrosion cell) is created and current flows from the anode to the pile. The anode is sacrificed to protect the cathodic metal (pile).
b) The _impressed current_ cathodic protection system uses anodes made of high-silicon iron or graphite that are energised by an external DC power source.

Numerous variations of these two procedures exist. The decision on which type of cathodic protection system to use is often difficult to make. Before a decision can be reached, a complete engineering and economic analysis of the problem must be made. Listed below are some of the advantages and disadvantages of the two types of system:

a) _Galvanic Anode System_

- Advantages
  - no external power source is required
  - no further adjustment is required after proper current drain is determined

Figure III.2
REPAIRING CRACKS BY EPOXY INJECTIONS

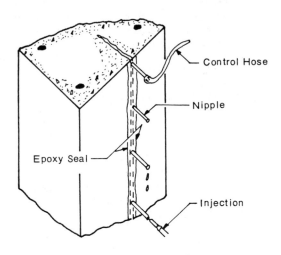

- installation is simple
- danger of cathodic interference is minimal
- negligible maintenance is required during the life of the anode
- current can be evenly distributed over a long structure
- over-protection at drainage points is minimised
- system costs are usually easy to estimate
- Disadvantages
  - current output is limited
  - anodes cannot be economically used in many highly resistive media
  - an excessive number of anodes and amount of material may be required to protect a large structure

b) <u>Impressed Current System</u>

- Advantages
  - system can be applied over a wide range of voltage
  - system can be designed for small or large current requirements
  - an extensive area of metal can be protected by one installation
  - applied voltage and current output can be varied over a wide range to meet changing conditions
  - current drain can be read easily at the rectifier
- Disadvantages
  - system must be carefully designed to avoid cathodic interference current from other structures
  - system is subject to power cuts
  - electrical inspection and maintenance are required
  - estimation of costs is complex because of the numerous variations in design.

Protection Against Scour

Scour of the substructure support is one of the more frequent factors that may cause or lead to structural failure or foundation distress. Scour is the removal of stream bed, backfill, slopes or other supporting material by stream, tidal action, dredging, propeller backwash, etc. The degree of damage depends on such factors as the character of the stream bed, the volume of water, and the shape of the structure.

If the cause of scour can be identified, such as change in the alignment of the stream, an inadequate waterway, or the presence of debris, the repairs made to scoured areas are much more likely to be successful. Determining the most effective solution to a scour problem is difficult. It is advisable to consult experts before undertaking the solution of a serious repair problem.

Spur dikes, jetties, deflectors, and other devices may be constructed to direct water away from a fill, bridge pier, or abutment. Caution is needed because only correctly designed and constructed training works are helpful in controlling scour and erosion. Repair of damage caused by channel scour may be as simple as the replacement of displaced material, or it may involve redesign of the footing, construction of training works or sheet piling, or other modifications of the structure or channel.

At sites where soil erosion has occurred because of stream or tidal action, it is common practice to place rock or riprap material in the void or to protect the replaced soil with riprap, bagged concrete riprap, or grouted or wire enclosed rock. Piers and abutments may be protected or repaired by placing sheet piling to keep material in place or to prevent further scour. Sheet piling should be driven to a depth where non-erodible soil conditions or rock exist. The overhead clearance required for driving under substructures may be a major disadvantage of using sheet piling. If supporting material has been removed from under a large area of the footing, consideration should be given to redesigning the foundation, including filling the void with concrete. In some cases, the footing may be extended by using sheet piling as forms for the extension and as stay-in-place protection against further scour. If scour has exposed supporting piles, it may be necessary, particularly if they are short, to drive supplement piles that are part of the extended footing.

III.3.2 Examples of repair work in Member countries

Some of the repair work carried out is described in more detail below:

- Foundation of a masonry bridge on timber piles in fast flowing waterway. This is the classical case of scour. If the piles still have sufficient penetration depth to guarantee satisfactory stability then as a protective measure a **monolitic foundation** can be restored. First a cofferdam of concrete filled bags, sometimes spiked together with reinforcement bars, is built around the piles and then the interior of the cofferdam filled with tremi concrete. Scour of silty river beds have sometimes required underpinning of the foundation.
- Erosion problems: Stone riprap is placed on a mattress at or beneath the channel bed level. The weight of the mattress preferably should be not less than 150 kg/m2. The slope of the protective riprap should be between 1 in 3 and 1 in 3.5. Heavier stones should be used for riprap, should it be necessary to use a steeper slope.

- Foundation on soft rocks subject to erosion: Reinforced concrete curtain walls, enclosing the footing or piles, are brought down at the uphill as well as at the downhill side.
- Increasing the bearing capacity of soil by injecting cement or chemical grout: Care has to be taken that grout pressure does not exceed existing ground pressure, otherwise more harm can be done than good.
- Rock or ground anchors: Rock or ground anchors are often used in abutments where the protective slope had to be removed (widening a navigable channel, road, etc.). Normally a protective sheet pile wall will be driven first in the case of ground anchors. Design and execution of rock and ground anchors requires great care and should take into account all factors likely to affect the bearing capacity and durability of the anchoring system. Most rock or ground anchors used nowadays are of the prestressed type. Provisions should be made to check periodically a representative number of anchors for possible loss of prestressing.
- Strengthening of an old river pier on timber piles: Widening and deepening of a navigable channel required the strengthening of the existing river piers. A steel pile wall was driven along the sides of the existing concrete footing. Since driving had to be done under the existing bridge, the piles -HZ-piles interlocked- were driven in shorter segments and welded in place. A second sheet pile wall was driven in an elliptical form to provide for a protection against ship collision and for a better streamlined cross-section. The space between the two walls was filled with tremi concrete. A reinforced concrete wall was poured and connected by four cross-beams. Two of the cross-beams are located at the centre lines of the bearings of the existing bridge. These two beams were bonded to the existing pier shaft by prestressing cables. Part of the reactions were transmitted by the prestressed cross-beams to the new sheet pile walls.
- Extension of existing foundations: This could be necessary in two principally different cases: (i) widening of an existing bridge and (ii) bridge replacement in the course of which the new bridge is constructed beside the existing one, traffic transferred to the structure, the old bridge dismantled and the new bridge moved into its final position.
    i) Widening of an existing bridge could be done in two different ways: if the existing piers have sufficient load bearing capacity, the new wider superstructure may be supported on them and on widened abutments. Otherwise the new widened superstructure has to be supported on extended piers and on extended abutments.
    ii) In the case of bridge replacement, it is normally necessary to strengthen and to extend the existing foundations for the new permanent superstructure in its final position. Preliminary piers and abutments are constructed in the preliminary position. Diaphragms of sufficient rigidity and load carrying capacity to support the loads of the final structure during the moving operation, connect the preliminary substructures with the final substructures. In all the cases mentioned above, the problem of possible differential settlements between the old foundations and new ones has to be given due attention.
- Rehabilitation of existing piles in water(6): Encasement in concrete may be done by a fabric form process. Fabricform jacketing involves placing special fabric around a pile and filling it with a concrete (grout or mortar). The piles have to be thoroughly cleaned of barnacles and rust and/or the deteriorated old concrete

must be chipped off. A light gage, split steel form is then assembled around the pile and forced into foundation mud to the required depth. Mud and debris has to be removed from between the form and the pile and then to be filled with structural grout to establish a seal and anchor the "can". Reinforcing steel is prefabricated in cages and assembled around the pile. Spacers have to be installed to hold the fabric jackets at a uniform distance from the pile. With the reinforcement and spacers in place, a precut, zippered fabric-form nylon jacket is wrapped around the pile and zipped shut. The lower end of the jacket is strapped to the outside of the steel can, the upper end is fastened to a steel hoop. Grout is introduced into the form by grout hoses inserted to the bottom of the jacket. The grout hoses are slowly withdrawn as the concrete inflates and fills the fabric form. Proportions and properties of the fluid structural mortar in a particular case are given in the Table below as an illustration. This technique is limited to water depths of approximately 10 metres.

Table III.4

EXAMPLE OF PROPERTIES OF FLUID STRUCTURAL MORTAR

Material	kg/m3
Portland cement	693
Fly ash or similar product	123
Sand	1087
Grout fluidifier ("intrusion aid")	6.5
Water	334

The abrasion resistance of fabric formed mortar is said to be superior to conventional concrete. The compressive strength of concrete cast in a nylon fibre fabric is higher than that of concrete cast in conventional watertight moulds as excess water is squeezed through the fabric.

III.3.3 <u>Splicing new steel sections to H-piles under or above water</u>(5)

This method is used to increase the structural strength of a deteriorated section of H pile when pile replacement is not practical. It should be considered as short-term repair unless the pile is jacketed.

The new steel sections are preferably channel sections about 45 cm longer than the distance between the extreme limits of the deteriorated section. The area to which the channel is bolted must be thoroughly cleaned. The principle of the method is shown in Figure III.3.

Figure III.3
SPLICING OF H-PILES

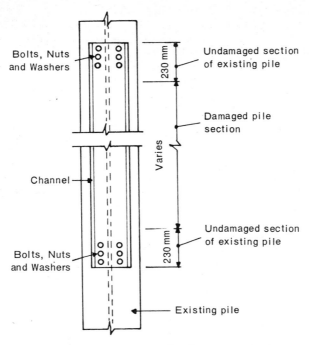

ELEVATION VIEW H PILE

III.4 MASONRY STRUCTURES

Reference is made to the OECD Report "Bridge maintenance"(1), sub-chapter III.3 Special Techniques for Maintenance of Masonry Bridges, in which typical defects of masonry bridges and special maintenance operations are described in more detail.

III.4.1 General

Many existing masonry bridges are now considered historical landmarks and part of the national heritage. Strengthening and widening will therefore often require that the appearance of the old bridge be kept completely unchanged or that the strengthened and widened bridge matches the general appearance of the old structure. In the first case, widening will usually not be possible whereas some form of strengthening work can often be done. In the latter case, many bridges have been strengthened and widened, i.e. adopted to modern traffic needs, and creating at the same time a structure of pleasant appearance. Strengthening and widening an existing masonry structure is a very delicate task both from a structural and an architectural point of view. Bridge engineers are well advised to seek the advice from experts in these fields. In almost all cases of strengthening and widening, special maintenance operations of the kind described in more detail in (1) will be necessary.

Decisions on strengthening an arch require an accepted method of assessment. At present two approaches are commonly used, based either on elastic theory or a non-linear collapse mechanism. Each requires the thickness of the arch ring to be known but it is rarely as indicated on the fascia which is often an architectural feature; even when construction drawings are available, masonry is notorious for not being constructed as shown on the drawings. A thorough assessment of the whole structure is needed as the primary cause of failure of arch bridges is associated with the foundations.

III.4.2 Foundations

The foundations are possibly the most important item of the structure as more failures have been caused by undermining than anything else during sudden flooding and periods of high flow. This is often intensified by river training works increasing the depth and straightness of the river. Thus considerable effort should be put into ascertaining the condition of the foundations. Specialist services, such as underwater television, or even divers may be required to probe between stones of the base to ensure that the interior is still sound. Unless the reason for undermining is defined and corrected, rehabilitation will be of little or at least limited use. Thus, rehabilitation may consist of replacing washed out material, e.g. by tremied concrete or by grouted stone fills, but individual circumstances may also require particular training works or protective measures to prevent recurrence. Excessive erosion of the pier base by aggressive streams (either mechanically by stones carried in flood, or chemically) may require the construction of a defensive outwater. Caution may be needed in restricted situations because this could aggravate problems of undermining. If the bearing capacity of the ground is marginal or if it is questionable whether the foundations will accept the extra load or the changed loading condition due to strengthening or widening the superstructure, pin piles, sheet piles or an enlargement to the abutment can be installed to lower the ground stress.

III.4.3 Arches

The arch ring can be strengthened in two ways - by adding material to the intrados or to the extrados. Adding to the intrados causes the least disturbance but is more difficult to complete successfully. Also it results in a reduction in headroom which is often restricted and will in most cases be the cause of new damage to the intrados as experienced on many bridges even where the headroom satisfies legal limits. Extra material may be placed by shuttering and pumping concrete (which is difficult to compact at the crown) or by fixing a mesh to the intrados and spraying concrete. In both cases, any shrinkage of the new concrete will tend to make the old and the new material separate radially. Also these impervious rings prevent natural drainage between the stones or brick work of the arch so that especial provision must be made to deal with water or under severe climatic conditions, such as in mountainous regions, with ice. Sprayed-on concrete will in any case change the appearance of an arch constructed of stone, brick or a combination of the two.

A more effective, but sometimes a more expensive, treatment is to remove the fill and cast the extra required thickness on the extrados of the arch. Usually a full ring is cast but occasionally only the end quarters are strengthened to act as cantilevers and reduce the effective span of the arch. Normal concrete placing techniques are satisfactory. Replacement backfill may be normal or lightweight concrete. The latter will reduce dead load on the foundations but may also reduce the factor of safety for stability of the arch.

Another expedient which is satisfactory where the increase in load carrying capacity is relatively small, especially for small span bridges, is to spread the wheel loads by casting a relieving slab at road level to act as an auxiliary bridge.

III.4.4 Widening

Widening of the road over a bridge without changing the structural system is possible if certain precautions are taken. The closeness of wheels to the spandrel walls results in horizontal forces which usually are not dissipated in the fill and arch ring. This force may be sufficient to displace the wall unless it is tied to the opposite wall. The spreading effect also operates on the wing walls and these must not be neglected. At present there appears to exist no entirely satisfactory method of preventing corrosion in the ties.

Some existing bridges have been strengthened and widened by changing the structural system whereby the appearance was kept mainly unchanged. Such a technique would normally be restricted to bridges of moderate span and length. Further, the geometry of the piers must be such as to accommodate appropriate bearing seats. The new system will usually be a haunched plate girder bridge and will require new abutments. This is normally only economic when the architectural merit of the bridge is a dominating factor.

III.5 CONCRETE STRUCTURES

Reinforced and prestressed concrete bridges in Member countries require rehabilitation and/or strengthening at an increasing rate. The reason for this situation is the ageing of the existing bridge stock and changing traffic requirements.

III.5.1 Scope of the problem

Need for repair usually arises from the following causes:

- spalling due to bursting pressures generated by corroding reinforcement,
- cracking due to loading including thermal effects, differential settlement, etc. (active cracks),
- cracking due to shrinkage, in-built weaknesses, unsatisfactory workmanship, etc. (dormant cracks),
- scaling of concrete due to freezing and thawing as well as wetting and drying cycles,
- accidental impact damage,
- disintegration of concrete due to the effects of aggressive water, alkali aggregate reactions, chemical attack, poor workmanship (porous or impervious concrete), or unsatisfactory compaction (as in areas of dense reinforcement),
- chloride and/or otherwise chemically contaminated concrete.

The above list is not comprehensive but includes most items likely to cause the more common repair needs. Surface repairs, especially techniques providing protection to the concrete surfaces, and internal repairs with regard to the injection of cracks and cavities are described in more detail in (1) and cover what could be called necessary and preventive maintenance.

Repairs in order to strengthen a structure are usually needed for reasons such as:

- to rehabilitate insufficient reinforcement due to:
  - erroneous analysis or design,
  - errors of construction,
  - corrosion of existing reinforcement,
  - elimination of the effects of redistribution of forces or stress.
- to avoid, after their injection, the reopening of important cracks and thus possible overstressing and fatigue damage.
- to transform an isostatic structure into a hyperstatic structure, e.g. by eliminating joints of cantilever girders.
- to increase the load carrying capacity of an existing bridge.

III.5.2 Repair and strengthening strategy

The best strategy can only be determined in the light of a thorough diagnosis of the causes of deterioration, faults and weaknesses and an assessment of current condition. <u>Wherever possible root causes should be corrected before repair is undertaken. Repair and strengthening operations should be mechanically and chemically compatible with the original material properties of concrete and steel and also with the basic structural concept.</u> Factors such as access, the need and duration of lane closures, and atmospheric conditions (temperature, moisture etc.) have a strong influence on choice of materials and methods.

III.5.3 Chloride contaminated concrete

Repair of chloride contaminated concrete remains essentially an unresolved problem even when the immediate causes can be corrected by e.g. waterproofing and drainage. This is especially so in those structures in which chloride contaminated concrete is part of the main load carrying structural components, e.g. box girder bridges. Ideally all contaminated concrete should be removed, corroded steel cleaned or replaced and then protected by good quality concrete or by an impermeable material. In practice it will seldom be possible to remove all contaminated concrete, e.g. in post-tensioned transversely prestressed concrete bridge decks. Moreover, the long-term effects of surface coatings to slow down corrosion by limiting ingress of moisture and oxygen are not yet fully explored. It is a deplorable situation that long-term piecemeal repairs often result from chloride contaminate and in extreme cases the costly alternative of replacement is sometimes necessary.

III.5.4 Cathodic protection of reinforcing steel in concrete decks(7)

Cathodic protection is a system for preventing the reinforcing steel from corroding. Although the same basic principles as described in III.3.1 are used in the case of concrete decks, the technology is different as indicated below.

Cathodic protection can be applied to a new deck or to a repaired deck. It lowers the potential of the steel reinforcement to the point where the anodic reaction cannot occur and corrosion stops. In the case of a damaged deck, all spalls and delaminations must be cut out and the damaged areas repaired. Concrete is a poor conductor of electricity. In order to get an even distribution of electrical power across the entire deck area, it is necessary to have an electrically conductive bituminous concrete layer.

Corrosion resistant anodes (made of a high silicon content cast iron cast in a flat disc shape, about 300 mm in diameter and 37 mm thick or a graphite disc about 250 mm in diameter and 37 mm thick) are introduced into this mix at several points. The direct current (DC) is provided by a rectifier which takes the 115 V AC or 250 V AC and rectifies it to produce a DC of the required voltage. The flow of the current is from the rectifier to the anodes and into the conductive mix which distributes the power evenly over the deck and the current then flows down through the concrete to the steel reinforcement and back to the rectifier. The current is controlled at the required voltage by a zinc-zinc sulphate half-cell buried in the deck close to a steel reinforcement. The principle of this circuit is shown below.

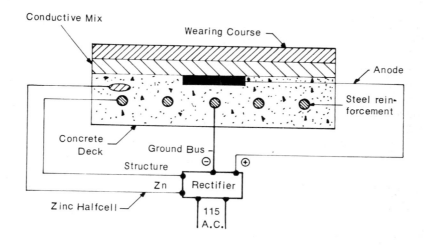

Figure III.4

PRINCIPLE OF CATHODIC PROTECTION OF REINFORCING STEEL IN CONCRETE BRIDGE DECKS

The circuit is a relatively "high" current and "low" voltage circuit (approx. 1.3 V and 2 to 3 A). The polarised charge must be within the range - 0.8 to - 1.25 V. Voltages numerically less than - 0.8 V may not provide protection and greater, numerically, than - 1.25 V may cause disbonding over a long period of time.

The cathodic protection system must be designed for each deck. It is important that the anodes be distributed along and across the deck so that there is a uniform distribution of power:

- for bridge decks 10 m or less in width, curb to curb, a single line of anodes may be used down the centre of the deck.

- for bridge decks 10 m to 21 m in width, two lines of anodes should be used along the deck, each line spaced 3 m from each curb.
- for bridges greater in width than 21 m, three lines of anodes should be used, one line down the centre of the deck and two lines of anodes, each spaced 3 m from each curb.
- the spacing of the anodes from each other in each line must not exceed 7.5 m. The end anode in each line must not be more than 3 m from the end of the deck. If the deck is in sections separated by expansion or fixed joints, the end anode must not be more than 3 m from each joint in each section.

It is important that all reinforcing steel in the deck is electrically continuous. For this reason, several "grounding" locations are chosen at random on the deck and connections are made to the steel, these are then connected to a common ground bus which also connects together sections of the deck separated by expansion joints.

Electrically conductive bituminous concrete is a hot mix consisting of stone, sand, coke breeze and asphalt cement. The presence of about 40 per cent to 45 per cent coke breeze makes the mix electrically conductive. It must be designed to suit aggregates and coke used at each location.

Lastly it should be pointed out that the same method can be used with a different technology to protect the reinforcing steel of other parts of the structure.

III.5.5 Replacement of corroded reinforcement steel

"Surface" reinforcing bars can be replaced by new bars provided concrete can safely be cut away to the necessary depth required for the replacement operation. In some countries when the material of the existing bar is weldable (a weldability test is recommended in any case) then the replacement bar can be directly welded to the existing reinforcement. This practice presents certain risks, especially in regard to fatigue. If the existing material is not weldable special anchors can be installed (glued in specially drilled holes) and the new reinforcement welded to the anchors. The new reinforcement should, in the latter case, overlap the sound existing reinforcement.

III.5.6 External reinforcement

Accurate diagnosis of causes of cracking is essential for effective repair. For example there is little point in injecting epoxy resin in corrosion induced cracks or active cracks because both types are likely to reopen. Active cracks in the vicinity of tendon couplings may result in high stress variations due to live load actions and may thus involve the hazard of premature fatigue fracture of the tendons.

Potentially dangerous cracked zones can successfully be rehabilitated or even strengthened by the direct approach of bonding external reinforcing plate to the concrete concerned. External reinforcing plates may be either steel plates or reinforced concrete plates. Usually steel plates are preferred for economical reasons and ease of operations. Bonding between steel and concrete seems to have reached a satisfactory state as regards suitable bonding materials and application practices. Long-term effects (far exceeding 10 years) cannot, however, be predicted with a high degree of accuracy

An average failure bond stress between steel and concrete was determined as 4.5 N/mm2 by Bresson(8), and the fatigue strength is given by L'Hermite(9) as 1.5 N/mm2 for a bond length of 200 mm.

Tests indicate(10) that bond lengths of approximately 300 mm are sufficient in order to develop ultimate load. Adhesion is due essentially to molecular forces of very small range of action whereas mechanical interlock is only of minor importance. Surface treatment prior to bonding should consist of a thorough cleaning of the surfaces concerned (surface films of low strength must be removed) instead of deliberately roughening the surface. Bonded sides of steel plates should preferably be sand-blasted and completely cleaned of any dust or grease.

There are different opinions on the optimum thicknesses of the bonding material and of the steel plate. Bonding material (usually based on epoxy resin) has been applied in thicknesses ranging from 150 to 300 microns up to approximately 0.5 to 5 mm and steel plates from 3 mm to 8 mm thick.

Ideally, from the bridge engineers point of view, bonding material should have the following properties:

i) with regard to handling and setting:
- easy workability at ambient temperature ranges
- adjustable viscosity
- sufficient pot-time (time from mixing to application)
- insensitivity to errors in proportioning
- rapid setting irrespective of temperature and humidity
- different coloured components in order to check adequate mixing

ii) with regard to final properties:
- permanent bonding to concrete and steel even when exposed to heat and moisture
- low creep
- temperature resistance

In practice there is no material available which will fulfil all these requirements. Ready-to-use materials are now commercially available which can be adjusted to suit specific job requirements and can be expected to yield satisfactory results.

Great care should be exercised with bonding around the plate edges. The reinforcement plates should be arranged in a staggered position.

This technique is still subject to further improvement. Current research and development work may lead to changes in present practice.

III.5.7 Supplementary prestressing

Strengthening by supplementary prestressing needs to be done with care to avoid overstressing parts of a structure and the existing level of prestressing should be known before any such work is undertaken. It is also good practice to inject all cracks caused by bending and/or shear before or after additional prestressing is applied.

Additional tendons can basically be applied in two ways:

- as vertical or inclined tendons
- as straight or curved tendons in beams

Vertical or inclined tendons are normally used to strengthen resistance in shear. A typical arrangement of additional vertical tendons for strengthening existing shear reinforcement is shown below:

Figure III.5

SUPPLEMENTARY PRESTRESSING USING STRAIGHT TENDONS

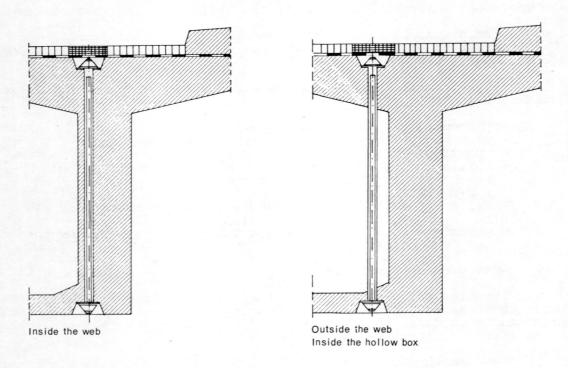

Inside the web

Outside the web
Inside the hollow box

The installation of longitudinal supplementary tendons in beams is the usual favoured method for strengthening resistance in bending. Three basic types are commonly used for their anchorage:

- end-anchorage at girder ends
- anchor blocks fixed to webs, to deck or bottom slabs
- anchorages in existing cross-members

End anchorages may involve casting a new anchor block at the girder ends and reconstruction of a new back wall in existing abutments as shown in Figure III.6. These operations may however necessitate large additional works on the abutments.

Figure III.6

END ANCHORAGE IN THE CASE OF LONGITUDINAL SUPPLEMENTARY PRESTRESSING OF BEAMS

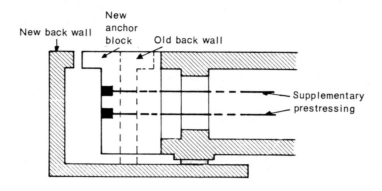

Additional anchor blocks can be fixed to existing elements using well-known and reliable techniques.

Anchorages in existing cross-members may be accomplished in two ways:

- direct anchorage in existing cross-members if they have sufficient strength to withstand the force induced by the supplementary prestressing.

Figure III.7

ANCHORAGE IN EXISTING CROSS-MEMBERS

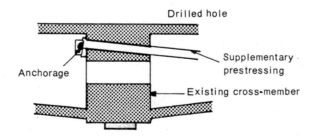

- anchorage in special constructed cross-elements if cross members have insufficient size or strength for a direct anchorage as shown in Figure III.8.

Figure III.8

ANCHORAGE IN SPECIAL CONSTRUCTED CROSS-ELEMENTS

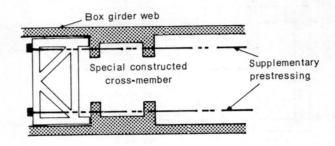

A three-span reinforced concrete bridge over the Tiber near Rome had to be strengthened in order to carry first category live loads (according to the Italian Code for bridge loads). The bridge was originally designed for second category live load. Strengthening involved adding 12 cm of new concrete to the webs of the main girder including additional longitudinal reinforcement and stirrups. It was accomplished by a prestressing arrangement that basically following the dead load bending moment curvature thus partially counteracting the dead load. Load tests made after completing the strengthening demonstrated very good agreement between calculated and actual values.

Two further examples, one in Germany and one in the Netherlands, are mentioned here in which structurally weak bridges have been successfully strengthened by supplementary prestressing. The two examples are not identical but have some similarity. The similarity in both examples is the development of a dangerous crack situation. In Germany, the coupling of tendons in the bottom plate of a box girder had been arranged in a line along a construction joint which developed into an active crack and the tendons finally failed in fatigue. In the Netherlands, joints of a segmental bridge with handed segments opened up and the width of the opening increased or decreased under traffic loads to such an extent that prestressing cables became visible. In both cases inherent structural weaknesses have been successfully remedied by supplementary prestressing.

The method applied in the Dutch example is described in more detail. The operation was done in the following steps:

- step 1: a temporary opening was cut in the bottom slab of the box girder in the centre segment of the span
- step 2: the cracked joints between girder segments were secured by steel plates
- step 3: cables in temporary location (1) were installed and stressed to 50 per cent of the full cable force to arrest the joint movement
- step 4: holes were drilled in the cross girders for final cables locations (2), (3) and (4)
- step 5: cables were installed in final locations (2) and (3) and encased in ducts
- step 6: the opened joints were injected

Figure III.9

DUTCH EXAMPLE OF SUPPLEMENTARY PRESTRESSING OF A BOXGIRDER

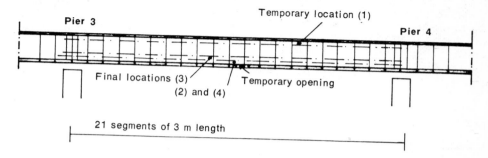

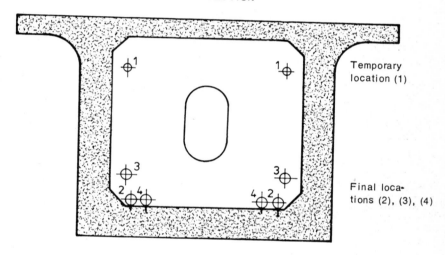

- step 7: after hardening of the injection material, cables in final location (2) and (3) were prestressed to 60 per cent to stop further joint movement
- step 8: cables in temporary location (1) were removed and installed in final location (4)
- step 9: cables in final location (4) were prestressed to 40 per cent
- step 10: all cableways were grouted
- step 11: the temporary opening was closed

Essentially similar operations as described above have been applied to several bridges in various countries.

III.6 IRON AND STEEL STRUCTURES

The majority of steel bridges have been built in the last fifty to 60 years. Many of the newer ones are of welded construction. The main rehabilitation problems arise through corrosion and especially in welded bridges fatigue can arise in components susceptible to resonant vibrations or to sufficiently high stress ranges generated by traffic. Vehicular impact is a frequent cause for rehabilitation. Inadequate safety margins or demands of increased traffic loadings may require strengthening that increases compressive collapse strength by the addition of stiffeners to flanges, webs and diaphragms.

III.6.1 Guide for defects in steel structures

Deterioration of Steel

- deterioration of the protective paint systems:
  - accumulation of debris
  - accumulation of moisture
  - flaking of the paint
  - cracks in the paint
- corrosion:
  - rust
    - light rust formation pitting the paint surface
    - moderate rust formation with scales or flakes forming
    - severe rust formation - stratified rust or rust scale with pitting of the metal surface

    A more detailed scale of rusting is given in a European Recommendation. (See DIN 53210 and ISO 4628/I - 1978)
  - electrolytic action - other metals that are in contact with steel may cause corrosion similar to rust
- chemical or physical attack:
  - air and moisture
  - animal wastes
  - deicing agents
  - industrial fumes, particularly hydrogen sulphide
  - seawater
  - welds where the flux is not neutralised

Abnormal Deformations or Movements

- abnormal vertical deflection
- abnormal horizontal deflection
- long-term deformation e.g. creep and sagging
- abnormal vibration due to traffic and/or wind
- excessive noise due to traffic
- excessive wear, due to traffic, of members accommodating movements such as pins
- buckling, kinking, warping and waviness:
  - due to overloading members in compression
- bent or twisted members:
  - due to vehicular impact

Fracture and Cracking

- fracture due to:
  - overloading
  - brittleness
  - stress corrosion
  - fatigue

- cracking:
    - due to sudden change in the cross-section of members
    - in welds in adjacent metal because of stress fluctuation or stress concentration

<u>Loose Bolts and Rivets</u> due to:

- overloads
- mechanical loosening
- excessive vibration

III.6.2 Iron bridges

This group includes both cast iron and wrought iron. In some Member countries there are a surprisingly large number of such bridges in use. Many of these bridges are now considered as part of the national heritage or attract the attention of preservation groups so that in many cases structural enhancement can only be carried out provided the appearance is unchanged.

Common faults requiring repair and/or strengthening are mainly concerned with corrosion and fractures of members having inadequate tensile strength to support the loads imposed by modern traffic.

Utmost care must be taken in all repair, rehabilitations and strengthening work not to mix different kinds of steel because of the risk of creating serious new corrosion problems.

The methods used for strengthening and rehabilitating iron bridges involve replacement of the structural material in such a manner as to produce a refurbished bridge but having its original external ironwork. The replacement can be steel or even reinforced or prestressed concrete.

Case histories in the United Kingdom include the following examples of strengthening schemes:

- a suspension bridge which had a prop inserted at the middle of its suspended span
- a suspension bridge which had new girders fitted which were strong enough to support the bridge on their own, so that the suspension system became redundant in a structural sense
- a box girder bridge which had an arch fitted to give added support to the original rail carriageway plus a second deck for road traffic.

III.6.3 Steel bridges

Deck Replacement of Older Steel Bridges

Many of the old bridges (usually truss or arch bridges) had either warped steel plates with a bituminous surfacing or a concrete deck. Due to insufficient waterproofing the steel plates were often corroded and many concrete decks were corroded. Several timber decks on **moveable** bridges had to be replaced mostly for economical reasons because they were too costly to maintain.

Bridge decks can be replaced by new concrete decks or by new orthotropic steel decks. Usually when a reduction in dead load or additional widening (adding cycle or pedestrian lanes) are necessary, replacement by an orthotropic steel deck is preferred. Bolting is the preferred method of connecting the new deck system to existing structural members.

Depending on the type of bridge and the load carrying capacity of its structural components, the new concrete deck is placed as a non-composite element, as a partially composite element (e.g. in composite action with the stringer and/or cross beams) or as a totally composite element (i.e. in composite action with all main load carrying elements).

The use of light weight concrete is often preferred in such cases where savings in dead load are an important factor. The use of aluminium decks, however, proved to be an unsuccessful method in Belgium and in the Netherlands. Sometimes to save weight, a type of steel grid decking is used where the grids can be left open or filled with concrete.

Strengthening of Structural Members

Strengthening usually involves more conventional techniques such as installing new diaphragms to existing double compression members (increasing buckling strength), strengthening or replacement of diagonals. Plate girders may be strengthened by external prestressing cables, anchored and fixed on the web in the required parabolic curvature acting in a similar way as in prestressed concrete.

Strengthening is sometimes concerned with compressive collapse and has involved the addition of stiffeners to flanges, webs, diaphragms and box girders.

Repair of Cracks

Fatigue can be due to any one or combination of the following reasons:
- poor detailing so that high stress concentrations are present
- increased traffic loading beyond what was anticipated by the designer
- an unexpected secondary structural action
- inadequate analysis of complex stresses
- a large undetected fabrication flaw.

Crack repair methods depend on the root cause of crack initiation. The structure should be analysed and especially those components which influence the overall safety of the structure.

Repairs can be made by techniques such as drilling holes at the crack tip (this should only be done in less sensitive locations), cutting out the cracked material and bolting cover plates in place, cutting out the crack and rewelding with a higher class weld (e.g. increasing the size and penetration of a fillet weld), strengthening the connection by introducing stiffening and by changing the structural action so that loads are supported in a way that prevents high stress ranges from developing.

Underwater Welding[11]

Arc welding has become an accepted procedure in underwater construction, salvage and repair operation. Underwater welds made on mild steel plate under test conditions at several United States Navy facilities have consistently developed over 80 per cent of the tensile strength and 50 per cent of the ductility of similar welds made in air. The reduction in ductility is caused by hardening due to drastic quenching action of the surrounding water. Structural-quality welds have been produced by means of special equipment and procedures that create small, dry atmosphere in which the welding is performed. However, this process is expensive.

Gas welding under water is not considered to be a feasible procedure.

A word of warning appears appropriate. Although arc welding and gas cutting are now common underwater techniques, electric shock is an ever present hazard. This hazard can only be minimised through the careful application of established procedures.

Strenghtening of old truss bridges by the use of a steel arch superposition scheme(12)

An interesting proposal has been made to strengthen old truss bridges. The strengthening scheme consists of superimposed arches, hangers, and additional floor beams. The concept of combining a truss with an arch is by no means a new system. Old covered wooden bridges, particularly the Burr Trusses, made extensive use of this system. The idea is that a light arch can carry a significant load if properly supported laterally. In this case, the truss provides the lateral support while the arch in combination with the hangers and additional floor beams provides the increased load carrying capacity. Additional floor beams and hangers are used for two reasons:

- the more uniform the load distribution the more efficient the arch becomes in carrying the load
- the floor systems of many old truss bridges are deteriorated and sometimes underdesigned and unreliable.

The principle of a steel arch superposition scheme is demonstrated in Figure III.10.

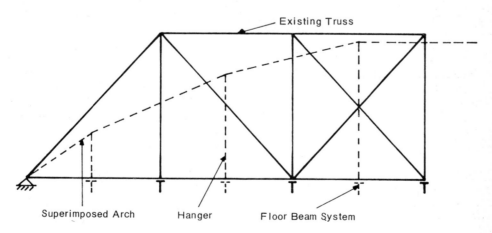

Figure III.10

STEEL ARCH SUPERIMPOSITION TO STRENGTHEN OLD TRUSS BRIDGE

The thrust of the arch can be resisted by one of the following means:

- the abutments, provided they are adequate and in good condition, or they can readily be repaired or strengthened
- a reinforced lower chord
- superimposed cables or rods
- properly designed and detailed stringers or floor slab.

The arch superposition scheme can be considered as an overall strengthening measure. The load carrying capacity of the entire structure is upgraded, thus allowing the live load to be increased. There is no need for temporary shoring or jacking for the installation of the superpositioned elements. The increase in dead load can be expected to be in the order of approximately 15 per cent to 20 per cent. The slender arch contributes only modest amounts of additional stiffness to the truss.

III.6.4 Suspension and cable stayed bridges

Cables may present a maintenance problem in both suspension and cable stayed bridges. The weakest part is around their anchorages and the lowest point in the curvature of suspension cables. Corrosion fatique can develop in individual wires to such an extent that it is necessary to replace stay cables in cable stayed bridges and, especially hanger cables in suspension bridges. Stay cables have been successfully replaced in some larger cable stayed bridges, for example in the Maracaibo Bridge and in a cable stayed bridge in Hamburg (Germany).

Main cables of suspension bridges are generally less of a problem but cases have been reported where corrosion of wires have necessitated major repairs. Strengthening and enhancement of suspension bridges may present problems because load carrying capacity is usually limited by the towers and suspension system.

III.7 COMPOSITE STRUCTURES AND MOVEABLE BRIDGES

Repairs, rehabilitation and strengthening methods are normally similar to those described in the foregoing sub-chapter. Special problems are briefly pointed out in the following.

III.7.1 Composite structures (composed of steel beams and concrete decks)

Comparatively few defects have been reported with well designed and fabricated shear connectors. Problems with concrete decks in composite structures are essentially of the same kind and order of magnitude as those found in concrete decks in regular structure. The same can be said for the main load carrying structural steel components.

Difficulties may be encountered with deck replacement or even major deck rehabilitation and strengthening operations in those composite bridges in which residual stresses have been introduced by sophisticated erection procedures combined with an elaborate casting sequence for the bridge deck.

III.7.2 Moveable bridges

Mechanical and electrical equipment and devices are beyond the scope of this report and will not be treated here.

Swing-span bridges and vertical-lift bridges have, apart from their machinery, essentially the same rehabilitation and strengthening problems as those mentioned in the foregoing sub-chapters.

In bascule bridges, especially with steel decks, surfacing may be a difficult problem. Reference is made here to the Danish experiences. In Denmark, either guss asphalt of the Danish type or a polymer modified bituminous mix is used.

III.8 COMPONENTS AND ACCESSORIES

The operational life of expansion joints and bearings is usually significantly shorter than that of the bridge and it is therefore necessary for designs to permit easy replacement. The strength and efficiency of bearings may be a limiting factor in situations where it is required to increase the load carrying capacity of a bridge.

Bearings or expansion joints showing signs of distress or unsatisfactory behaviour (e.g. due to unexpected or underestimated amount of rotation or movement, unsatisfactory waterproofing, etc.) are usually replaced by new bearings or expansion joints.

Large expansion joints, especially those of the roller-plate-type can sometimes be repaired satisfactorily. A satisfactory remedy could be as follows:

- replace all worn-out bushes
- prestress the sliding plates to the bridge and to the abutment by prestressed springs.

Special reference is made in this connection to the work of the P.I.A.R.C. Committee on Road Bridges. This committee works, inter alia, on problems related to waterproofing and surfacing, expansion joints and bridge bearings.

III.9 SEISMIC STRENGTHENING (RETROFITTING)

The collapse of a highway bridge during an earthquake will in many cases sever vital transportation routes at a time when they are most needed to provide emergency services to a stricken area.

There are four areas where local failure has a high likelihood of occurrence:

- bearings and expansion joints
- columns, piers and footings
- abutments
- liquefaction of foundation soil

Guidelines for the seismic retrofitting of existing highway bridges are being developed by the Applied Technology Council ATC-6-2 project in the United States(13). The project is expected to be completed in 1983. Special experience is also available in Italy (Friuli, 1976 and Compania-Basilicata, 1980) and has influenced design philosophy and rehabilitation requirements.

III.9.1 Bearings and expansion joints

In major earthquakes the loss of support at bearings has been responsible for several bridge failures. Although many of these failures resulted from permanent ground displacements, several were caused by vibration effects alone. Even relatively minor earthquakes have caused failure of anchor bolts, keeper bar bolts or welds, and nonductile concrete shear keys.

In the case of the differential displacements at expansion joints, elastic response spectrum analysis gives displacements that are often below those that are in line with observed bridge behaviour during past earthquakes. In addition to the non linear behaviour of expansion joints, possible independent movement of different parts of the substructure and out-of-phase movement of abutments and columns resulting from travelling surface wave motions also tend to result in larger displacements.

Certain structural configurations are exceptionally vulnerable to collapse in the event of a loss of support at the bearings. Such structures would be prime candidates for retrofitting. Simple or suspended spans in which no redundancy exists are particularly vulnerable. To a lesser degree, this is also true in the case of a structure with a small redundancy, such as continuous bridges with only one support between expansion joints.

Techniques for retrofitting bearings include:

- longitudinal joint restrainers
- transverse bearing restrainers
- vertical motion restrainers
- bearing seat extensions
- replacement of bearings
- special earthquake resistant bearings and devices.

III.9.2 Columns, piers and footings

During an earthquake, the interaction of the columns and piers with their footings will determine the probable mode of failure for these components. Plastic hinges may occur in the column end regions or at the footing. Piers can develop plastic hinges in the end regions about the weak axis only. The location of plastic hinging will dictate the modes of failure that should be investigated.

Four modes of column failure are considered in evaluating columns:

- shear failure in the column
- anchorage failure in the main longitudinal reinforcement
- flexural failure in the column due to inadequate transverse confinement
- failure of the splices in the main longitudinal reinforcement.

In general it is more difficult and less cost effective to specifically retrofit columns, piers and footings than it is bearings. A cost-effective retrofit measure can be achieved without the necessity for retrofitting the substructure if force-limiting bearing devices can be added between the superstructure and columns, piers, or abutments. This method is used in Italy and France and is proposed for use in New Zealand.

To date there are very few retrofit methods that have actually been tried on seismically deficient bridge columns. Several methods have been proposed such as:

- improved confinement
  - prestressing wire is wrapped under tension around the column or steel reinforcing that is prestressed on the outer face of the column through the use of a specially designed turnbuckle. In both cases the steel is protected with a layer of sprayed on concrete;
  - a solid steel shell that would be welded in place around an existing column;
  - steel banding of the type used for packaging materials (this method would be effective for smaller sizes only)
- reducing the maximum shear force on a column by decreasing the yield moment at one or both ends of the column. This can be done by cutting the longitudinal reinforcing bars. Since this will increase the ductility required at the points of flexural yielding, this technique must be employed with caution, but should be considered when columns are overreinforced for flexure resulting in little or no flexural yielding during an earthquake.

III.9.3 Abutments

Failure of abutments during earthquakes usually involves tilting or shifting of the abutment either due to seismic earth pressures or inertia forces transmitted from the bridge superstructure. Abutment tie back systems and settlement slabs are possible retrofitting techniques.

III.9.4 Liquefaction of foundation soil

Most foundation failures during earthquakes are the result of excessive soil movements such as occurs due to liquefaction. There are two approaches to retrofitting that will mitigate these types of failure:

- eliminate or improve soil conditions that tend to be responsible for seismic liquefaction
- increase the ability of the structure to withstand large relative displacements similar to those caused by liquefaction or large soil movement.

Some methods are available for stabilizing the soil at the site of the structure. Each method should be individually designed making use of established principles of soil mechanics to insure that the design is effective and that construction procedures will not damage the existing bridge. Possible methods for soil stabilization include:

- lowering of groundwater table
- consolidation of soil by vibrofloatation or sand compaction
- placement of permeable overburden
- soil grouting or chemical injection.

At a site subjected to excessive liquefaction, methods to improve the structure may be ineffective unless coupled with methods to stabilize the site.

III.10 STRUCTURAL ANALYSIS AND LOAD TESTING

In all Member countries, numerous older bridges are structurally deficient and, as a consequence, are posted for weight and speed restrictions or have been included in bridge strengthening or bridge replacement programmes. In many cases such structurally deficient bridges have inadequate load carrying capacity on the basis of analytical evaluations. Often, analytical evaluations are based on analytical models either too simplified, e.g. underestimating the degree of interaction among various components of a bridge, or neglecting the actual behaviour of the material used.

Intensive diagnostic testing of bridges(14) have shown that most bridges possess more strength than they are given credit for having. Further information on full-scale loading tests are given in a previous OECD Report(3).

Much progress has been made in developing advanced methods for structural analysis. In many countries, test programmes have been undertaken to determine the actual ultimate load carrying capacity, especially of concrete bridges, and compare the results with the predicted ultimate behaviour. Load carrying capacities well in excess of those predicted by conventional ultimate strength theory were often observed.

The non-linear finite element approach has shown good agreement with results obtained from tests to destruction and demonstrated its ability to predict the ultimate behaviour of structures even those of relatively complex design such as multi-cell box girders of straight, curved, and skew plan geometry(15).

The use of current codes, particularly when they are based on limit states, often shows that higher loads can be carried compared to older codes using working stress methods.

Reference is also made to a research done by the Task Group "Effects of imperfections of steel plated structures on their ultimate strength" of Committee II of IABSE.

In general, the cost benefit of rating older bridges employing advanced methods of structural analysis or diagnostic testing or a combination of both may result in a relaxation of weight restrictions and even the deferment of strengthening work or structure replacement for some time.

REFERENCES

1. OECD, ROAD RESEARCH. Bridge maintenance. OECD, Paris, 1981.

2. OECD, ROAD RESEARCH. Bridge inspection. OECD, Paris, 1976.

3. OECD, ROAD RESEARCH. Evaluation of load carrying capacity of bridges. OECD, Paris, 1979.

4. COMITE EURO-INTERNATIONAL DU BETON. Rapport préliminaire de la Commission IX "Comportement en service, entretien et réparations", Bulletin d'information No. 138, Paris, 1980.

5. TRANSPORTATION RESEARCH BOARD, National Research Council. Underwater inspection and repair of bridge substructures. National Co-operative Highway Research Program Synthesis of Highway Practice 88. Washington, D.C., 1981.

6. BINDHOFF, EW. and KING, JC. World's largest installation of fabric-formed pile jackets. Civil Engineering, ASCE, 1982.

7. ONTARIO MINISTRY OF TRANSPORTATION AND COMMUNICATIONS. Cathodic protection for concrete bridge deck restoration. 1981.

8. BRESSON, J. L'application du béton plaqué. AITBTP No. 349, 1977.

9. L'HERMITE, R. Use of bonding techniques for reinforcing concrete and masonry structures. Mat. et Contr. (RILEM) 10, 1977.

10. ROSTASY, FS, and RANISCH, EH. Verstärkung von Stahlbetonbauteilen durch angeklebte Bewehrung. Betonwerk + Fertigteil-Technik, Volumes 1 and 2. 1981.

11. TRANSPORTATION RESEARCH BOARD. National Research Council. Underwater inspection and repair of bridge substructures. Ref. 50 Underwater Cutting and Weldings. U.S. Naval Manual. Department of the Navy. National Co-operative Highway Research Program Synthesis of Highway Practice 88. Washington, D.C., 1981.

12. ANON. International Conference on Short and Medium Span Bridges. Proceedings, Volume 2 (pages 334-341). Toronto, 1982.

13. IABSE. Maintenance, repair and rehabilitation of bridges. Symposium, Introductory Report. Volume 38. Washington, D.C., 1982.

14. BAIDAR BAKHT, P. and CSAGOLY, F. Diagnostic testing of a bridge. Journal of the Structural Division, 1980.

15. ANON. International Conference on Short and Medium Span Bridges. Proceedings, Volume 1 (pages 20-34 and 392-405). Toronto, 1982.

Chapter IV

REHABILITATION AND STRENGTHENING POLICIES: PROPOSALS

IV.1 THE FUTURE CHALLENGES

IV.1.1 The need for prediction

Road networks may be regarded as permanent installations, and elements in a network - such as bridges - are generally constructed to satisfy this assumption of permanence. Bridges, however, deteriorate and society's needs and requirements in regard to established road networks change continuously. The former fact leads to rehabilitation, the latter to changes of existing bridges such as strengthening.

Rehabilitation and strengthening are also generally accomplished with the aim of meeting the required permanence. Regardless of whether the question is the construction of a new bridge or the rehabiliation on strengthening of an existing one, the work cannot be accomplished satisfactorily unless there is a reasonable understanding of society's needs and requirements with respect to road bridges, as well as what needs, requirements and changes the present and future bridge stock may impose upon bridge authorities.

IV.1.2 Future traffic demands

Since the appearance of the automobile the growth in traffic volume has exceeded all expectations, however, there are signs that this growth is levelling off and recent forecasts of traffic volume show either a decrease or a minor increase. Undoubtedly there will be local variations that have to be accounted for.

The allowable gross weight of vehicles and of axle loads have had a similarly strong upward trend during recent decades. This growth has had a major influence on live load assumptions in design codes for bridges and the increase has given rise to a broad demand and need for strengthening of bridges. The 1983 OECD report on "Impacts of Heavy Freight Vehicles" contains the following forecast:

"The tendency, common to all countries, is to upgrade fleets by increasing the number of vehicles at the top of the scale. The reason for this is economic. For a given road network, the cost per payload tonne transported falls with gross vehicle weight and the energy cost per payload tonne transported follows the same law."

The report also makes the following observation:

".....certain general trends are clearly common among OECD Member countries. First, road transport continues to increase its share of freight transport. Secondly, the truck fleets in Member countries continue to expand and to get heavier and larger, as the business community seeks the efficiencies of larger payloads. Further, the current trends toward tandem and tridem axles, as well as wide base tyres, are likely to affect future vehicle configuration. Member countries have also experienced steady increases in the average length of haul in road freight transport, as the heavier vehicles have become more and more competitive for long-distance movements of certain types of commodities, such as metal products, chemicals and manufactured goods."

Another important factor in traffic load assumptions is the number and size of exceptionally heavy vehicles (EHV). The general technological development in industrial nations has a continuous impact as developments in industrial technology override constraints such as bridges with restricted load carrying capacity. Some countries expect a continuous increase in the number and size of EHV's as well as an extension of the routes serving such vehicules. EHV's are expected increasingly to use minor roads in rural areas.

The future will undoubtedly bring increased demands for traffic safety in general. For bridges this will involve considerations such as

- demand for climatic countermeasures, e.g. measures to prevent early icing on bridges, especially steel bridges,
- increased structural capacity of crash barriers both along the parapets and around existing bridge columns;
- an increased ability for overpasses to withstand impacts from vehicles to guard against the many crashes between bridges and trucks loaded to excessive heights;
- strengthening of bridge piers and superstructures against possible ship collisions, since such collisions are likely to become more frequent.

IV.1.3 Environmental demands and considerations

Several OECD countries, notably Germany, Japan and the Netherlands, have installed noise barriers on road bridges situated near residential areas - a feature that was inconceivable in bridge building just ten years ago. This demand has created a new series of bridge problems such as increased wind loads, snow piling up on bridge decks hindering traffic, unexpected thermal loads due to the shadow effect of the barriers, inability to employ cantilever inspection vehicles and of course increased maintenance costs.

In New York City the Manhattan Bridge crossing the East River is at present undergoing a major rehabilitation at a cost of up to $100 million. The main cables are deteriorated to the extent that the original overall safety factor of 5 has decreased to around 2 and is decreasing further. A replacement of the main cables will cost an additional $125 million. A replacement of the bridge as such is out of the question because it has been deemed environmentally unacceptable to provide a right-of-way for a new structure adjacent to the present one.

In Germany, in Dusseldorf and near Koblenz, such a denial of replacement has been avoided, but at high additional construction costs. At both locations new bridges, spanning the Rhine, were constructed on temporary substructures adjacent to the existing structures. While the traffic was detoured temporarily over the new structures, the original structures with substructures were demolished. A second set of new substructures were then constructed and the new superstructures pushed horizontally into the alignment of the original old structures.

The above mentioned examples of non-technical, social and environmental restrictions may indicate future challenges for bridge authorities. What today are rare examples for such restrictions may tomorrow be common requirements.

IV.1.4 Technical restrictions

The development in bridge technology, and especially the increased ability to successfully design bridges with complex integrated static behaviour, creates problems with respect to rehabilitation works, because such modern integrated bridge structures may not be repaired and rehabilitated without permanent changes in the original static system. The changes may be so severe that replacements will be both technically and economicly more attractive than even limited rehabilitations, e.g. the high costs that are related to repairs and rehabilitation of prestressed structures can make replacements more economic when such replacements are acceptable with respect to traffic requirements.

Older design concepts, on the contrary, are often more suitable for rehabilitation and strengthening operations. The reason for this is the obvious far simpler static design concept that makes control over necessary interventions in the static system during rehabilitation and strengthening easier and safer.

IV.1.5 Socio-economic aspects

As a result of continual changes in public values, there appears to be an increasing willingness to pay the additional costs of environmental requirements, traffic safety or traffic serviceability. Thus, in many cases the direct costs of rehabilitation and strengthening may constitute a decreasing part of the overall costs. The willingness to pay these additional costs may be explained by the fact that the overall costs of constructing and maintaining road systems are declining relative to the increasing costs of social welfare, education and defence.

It is vital to attempt to foresee future demands and their costs and to evaluate these costs and their expected benefits to strike an acceptable balance between the many demands and the ability to meet them.

IV.1.6 The future role of strengthening and rehabilitation

The increased importance of strengthening and rehabilitation of bridges can be characterised by:

- increased loads and traffic on existing bridges,
- increased technical challenges in areas not yet fully envisaged,
- increased costs due to (1) limitations of the possibilities for replacement, (2) technical difficulties in rehabilitation and (3) high traffic demands,
- increased maintenance owing to the high cost of rehabilitation with both activities frustrating the demand for unrestricted traffic flow.

The above discussion highlights the most important future challenges. It should be stressed that some of these challenges will only be faced in densely populated and technically highly developed areas, and that the bridge problems taken as a whole will differ considerably from one area to the next. We can thus expect that a large number of bridges - probably the majority of the bridges - will not be faced with environmental demands or demands for increased traffic safety and will involve few technical restrictions.

Thus, the question posed is: how to formulate bridge policies to address this complex variety and future challenges?

IV.2 THE FORMATION OF A GENERAL POLICY

IV.2.1 Aim of a general policy

Each road agency has a hierarchy of planning activities starting with the overall agency policy descending in steps until it reaches the content of specific decision considerations. This report endeavours to give guidelines to decision considerations for bridge rehabilitation and strengthening as well as outlining a framework for a general policy for such activities.

The content of the individual decision considerations may vary in accordance with the size and importance of the bridge work in question. When rehabilitating or strengthening major bridges, elaborate analytical techniques for evaluation of the various solutions may be employed while an ordinary rehabilitation and strengthening work may be carried out in accordance with preset general guidelines. In all cases, however, it is important to formulate a general policy that ensures the available funds are allocated in accordance with the agency's overall objectives and policies.

A general policy must take a number of parameters into account - parameters that have been described in the preceding subchapters - and it should be operational for dealing with sophisticated cost-benefit analyses of major rehabilitation and strengthening works as well as for general guidelines of day-to-day operations.

The solutions - and their consequences - to be considered in each case may vary from no action, any degree of temporary action, full rehabilitation or strengthening to a replacement. In principle the decision could be reached through a cost-benefit analysis of the available options, however, such an approach may prove to be a laborious task in day-to-day operations and perhaps unjustified in the majority of cases.

In the following subparagraphs a simple, operational and robust management tool is introduced. The tool attempts to link the more important policy factors into a comprehensive framework for decision making.

IV.2.2 A division into policy factors

Bridges may be divided into elements with relatively short life spans such as pavements, waterproofings, expansion joints, paint etc., and bridge elements with long life spans such as structural elements in bridge decks, columns, foundations etc. Such a division between short life and long-life bridge elements is useful because of the stronger economic motivation to rehabilitate and strengthen long-life elements than short-life elements.

A division between ordinary bridges and major bridges is warranted when considering the differences in technology and economy.

Traffic considerations, and especially serviceability, lead to a division between bridges on rural roads, secondary highways, primary highways and motorways. Environmental restrictions may lead to a similar division.

The general policy for rehabilitation and strengthening of bridges may thus contain the following factors:

- Short life elements,
- Long life elements, ordinary bridges
- Long life elements, major bridges

* Bridges on motorways and primary highways
* Bridges on secondary highways
* Bridges on rural roads

The above six factors interact in a matrix form as indicated below.

IV.2.3 A management tool

The above-mentioned policy factors enter into decision considerations when evaluating possible bridge rehabilitation or strengthening strategies. To ensure that policy factors are fully assessed, it is adviseable to perform cost/benefit analyses. However, in the bridge sector it is extremely difficult to estimate future expenses and to determine benefits. An evaluation concept based on differing discount rates is therefore introduced hereunder.

In the OECD report on bridge maintenance the general economic considerations were discussed in some detail. The conclusion from this discussion was that economic considerations could be expressed by means of discount rates that determine the present value of future costs. It should be stressed that the use of discount rates in this report must not be mistaken for the use of discount rates by governmental budgetary authorities to control public investment policies.

The application of discount rates in bridge management aims to achieve optimum results within a given, limited economic framework by funneling resources into projects of recognised importance at the expense - though not the abandonment - of projects of lesser importance.

Major bridges are considered more important than ordinary ones and long-life elements are considered more important than short-life elements. Thus short-life elements may carry a high discount rate, long-life elements in ordinary bridges a medium discount rate and long-life elements in major bridges a low discount rate. A similar variation is applicable for the different types of roads. Rural roads may carry high discount rates and important highways and motorways low discount rates. In order to quantify and calibrate the varying discount rates it may be advisable to employ cost-benefit analysis on a number of representative cases.

With this background it is possible to make a proposal for the formation of a general policy - a policy matrix for rehabilitation and strengthening of bridges as shown on Figure IV.1 By assigning each policy factor a discount rate (e.g. secondary highways 2 per cent and short-life elements 5 per cent - net of inflation), the policy element may be given its actual

Figure IV.1

BAR GRAPH EQUATIONS ILLUSTRATING DISCOUNT RATES FOR THE POLICY ELE

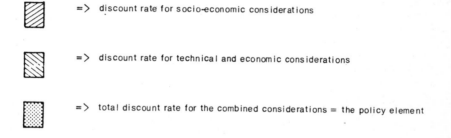

=> discount rate for socio-economic considerations

=> discount rate for technical and economic considerations

=> total discount rate for the combined considerations = the policy element

discount rate by adding the two rates (the policy element: short-life elements on secondary highways may then be given the discount rate = 2 per cent + 5 per cent = 7 per cent).

What is described qualitatively above is possibly no more than what is the normal practice for many bridge authorities, however, quantifying it may serve the purpose of providing a more comprehensive view of a functioning general policy.

In the following subchapter the various parameters that the policy contains are discussed. It is, however, important to stress that technical considerations in rehabilitation and strengthening are closely interrelated with economic considerations.

For each case there may be a variety of technical solutions that will solve the immediate problem, but the solutions may differ in qualities such as durability, appearance, replaceability and serviceability. In general, the technical considerations must take these parameters into account while pursuing an acceptable quality at the lowest price.

IV.3 POLICY PARAMETERS

IV.3.1 Future increase in functional demands

The OECD report on "Impacts of Heavy Freight Vehicles" outlines the trends in road freight transport and heavy truck usage as follows:

"Road transport has consistently increased its share of the freight market. In the process, the trucking industry continues to grow, and truck fleets in Member countries have continued to become bigger and heavier, as the business community seeks the efficiencies of larger payloads."

The 1979 OECD report on "Evaluation of Load Carrying Capacity of Bridges" deals with the increasing problem of overload:

".... gross vehicle weights and axle loads must not exceed the national statutory limits .... . The Group has, however, noted from the results of an enquiry conducted that <u>in some areas</u> up to 40 per cent of freight vehicles do in fact exceed these limits and this, to an overwhelming extent, without legal permits. Maximum overloads as ascertained in some countries and expressed as a percentage of maximum permissible loads may be as high as 100 per cent for vehicles with two axles whereas overloads commonly observed for road trains may be up to 20 per cent ..... .

It should be noted that systematic control of actual axle loads of freight vehicles should be part of normal motorway and road management programmes in OECD Member countries."

These trends have only a limited effect on bridges designed and constructed in accordance with modern live load requirements. However, OECD countries have a large number of older bridges that have been designed for quite different and much lighter loads.

Bridges constructed before about 1960 have load carrying capacities that often are close to or under the loads of the present day largest trucks, e.g. in the United States more than 25 per cent of all bridges are reported to require load restrictions. An increase in the allowable gross weight and axle load may drastically increase the number of load restricted bridges and thus create numerous obstacles to the free flow of traffic, especially on rural roads.

Enforcement of load restrictions is, however, increasingly difficult. The United States is now attempting to enforce load restrictions more vigorously whereas Denmark has temporarily abandoned prosecution of overload violations because of increased requirements to legal evidence.

There may be a further growth in the number and size of EHV's. Traffic load assumptions in design codes in the various countries attempt to foresee this development. Most OECD countries have established heavy-load grids, where the load carrying capacity of the bridges attract special attention. In some countries the problems associated with the administration of EHV's have been dealt with satisfactorily (United Kingdom). The United States is undertaking a thorough study of the problem by several committees and researchers, and in the Netherlands similar studies are going on.

Outside the heavy load grids there are a significant number of bridges not rated. The lack of rating is generally considered an important problem but no accurate estimates as to the size of the problem has yet been made. In Japan roughly 20 per cent of the total bridge stock is not rated. In the United States as many as 60,000 bridges outside the Federal-aid highway system must be evaluated without original design computations or as constructed plans.

The maximum allowable gross weights of EHV's including weight of trailers in the established heavy load grids vary from 300 to 500 tonnes.

Allowable maximum gross weights are dependent on a number of factors including the wheel configuration of the trailer and its suspension as well as the load equalizing devices between the trailer wheels. It should be stressed that further technical development of heavy-load trailers is as important to the protection of bridges as is the strengthening of bridges.

Widening of bridges to accommodate increased traffic is under way in all countries. The operation poses special problems that are considered in Chapter III. Above all it is usually an expensive operation and it is often rejected in favour of replacement by a new bridge. Narrow bridges are often traffic hazards and widening operations are, therefore, expected to upgrade overall bridge serviceability.

The clearance under bridges is an important requirement that causes some problems. The design assumptions for the required clearance under bridges vary from country to country. The minimum seems to be 4.2 metres and for most motorways on the European continent it is 4.6 metres. In the United Kingdom and in the United States the design clearance is 16 feet equal to 4.9 metres on trunk highways and 14 feet (4.3 metres) on less important roads.

It appears to be an accepted fact internationally that the allowable height of vehicles must be limited to 4 metres and it is likely that this limit will be maintained if only because many roads in the OECD countries cannot sustain higher vehicles.

Although it is not expected that the allowable height of trucks will increase above 4 metres, the problem of exceptionally large or tall shipments still remains. As such shipments can not be moved on motorways, the general practice is to use adjacent older highways with few or no overpasses, but traffic safety considerations may tend to reduce the number of level crossings replacing them with new bridges and overpasses and thus introducing new limitations for tall shipments.

IV.3.2 Environmental impact

The developed nations have established regulations in recent years to reduce damage to environmental and historical assets.

Where and what kind of bridges pose potential environmental and preservation problems?

- bridges crossing water. Any action below water level may prove to be a problem. In streams or rivers any change of riverbed, vegetation or material may be disruptive to the natural environment. Adjacent work may pollute the water too, e.g. drainage around abutments may contribute to water pollution.
- historic bridges for obvious reasons. It is advisable for a bridge authority to seek an overall agreement with the preservation authorities covering historic bridges because of the complexities and ambiguities involved in the designation of historic bridges.
- bridges in congested areas. Any change including minor repair works may create an opportunity to question the justification of the bridge (and the road) as such.

IV.3.3 Description of possible alterations

The main concern when discussing strengthening of bridges lies evidently with the load carrying ability to sustain present traffic. In all OECD countries there are a large number of posted bridges, i.e. with an insufficient load carrying capacity. There are also a number of bridges that are not posted but nevertheless have insufficient load carrying capacity.

There can be three major reasons for the restricted carrying capacity: (1) the bridge was designed for a load capacity that is now insufficient for the present use, (2) design or construction was deficient or (3) structural elements have deteriorated over time. In general, lack of load carrying capacity requires alterations of the structural system, mainly in the superstructure.

The techniques applied for the various types of bridges have been described in Chapter III of this report and shall therefore not be repeated here. There is a need for further development and research to improve the present techniques for repairing and strengthening of bridges. These are discussed in Chapter V.

From a technical and economic point of view, there are limitations to the extent of possible alterations of bridges.

Other forms of strengthening that do not stem from lack of load carrying capacity can be aimed at securing a higher resistance against impacts from vehicles or ships to the substructure or the superstructure. A large number of older bridges require strengthening of crash barriers which may also require structural changes in the superstructure if it is unable to absorb the increased impact forces. Old through trusses are especially vulnerable to impact damage from vehicles.

A special, very complex and expensive operation is the strengthening of bridge foundations, which may be caused by errors in design and construction. Also, changes to the immediate environment, such as a drop of water table for bridges founded on wooden piles, will call for this kind of strengthening.

IV.3.4 Considerations regarding rehabilitation and strengthening versus replacement

The obvious alternative to rehabilitation and strengthening operations is the replacement of a bridge. Replacements in this context cover either a replacement of the total structure or major parts of a structure such as replacement of a bridge deck. There are a number of factors that influence the choice of action, some of which may be decisive in particular cases. The more difficult of these factors are demands that spring from third parties and sometimes also political sources, who may not give sufficient attention to the magnitude of the resources needed to satisfy their demands.

The preservation of historical monuments, valued resorts and landscapes or even an infringement of such treasures are examples of compulsory demands on bridge authorities. Other environmental considerations such as noise and water pollution may be more manageable, if the requirements are well known and legally specified.

For more clear-cut technical/economic aspects of the decision of strengthening and/or rehabilitation versus replacement, the possible alternative solutions include:

- no immediate action plus various numbers of subsequent actions,
- any degree of immediate temporary action plus subsequent actions,
- full rehabilitation and/or strengthening,
- replacement.

By discounting the costs of the actions in the analysis, all possible solutions may be presented by their total present costs and the most inexpensive solution can be determined.

In connection with this analysis it is important to point out some problems related to rehabilitation and strengthening:

- work and materials needed can very often not be reliably assessed and, as a result, rehabilitation work is often more extensive than anticipated in specifications; claims for extra work are generally experienced,
- the mix of old and new makes it difficult to guarantee work,
- evaluating the lifespan of a rehabilitated or strengthened bridge is difficult.

These aspects may favour replacements and should be accounted for.

There may be some overriding concerns such as unacceptable risks of collapse or impediment to traffic which may exclude otherwise attractive solutions, however, the procedure described here will in the majority of cases be a valuable basis for a decision.

IV.4 ADAPTING STRUCTURES TO EMERGING NEEDS

IV.4.1 Evaluation of bridge structures

In the last decade, bridge authorities have realised the necessity of bridge inspection. The OECD report on bridge inspection was the first important international work in this field, and it has served as a model for bridge authorities in many OECD countries. The bridge inspection schemes described in this report create the fundamental information and background for rehabilitation of bridges. The inspection data have proved to be valuable technical information that can be used for purposes such as assessments of overall conditions and technical research and development. This has led to computerised data bank systems in many countries, where all the valuable information is stored and easily accessible. Systems of this kind have been described in the OECD report on bridge maintenance.

In the OECD report on evaluation of load carrying capacity of bridges there is an extensive description of the background for evaluations and analyses of bridges, which is highly relevant when treating problems of bridge strengthening. The report also describes existing rating systems that aim to give a systematic and overall assessment of bridges on the road networks. The rating systems in various countries differ considerably and this includes the treatment of essential items for bridge rating such as load, stresses, fatigue considerations, assumed structural behaviour and calculation methods. The report recommends that comprehensive bridge rating systems should be established and emphasises the need for closer international co-operation in this field.

While important data are compiled and sophisticated data analyses carried on the basis of frequent bridge inspections to aid the rehabilitation of bridges, there appears to be a need for a similar system for evaluation of load carrying capacity of bridges. There are two main problems in this respect. Since there exists a large number of older bridge structures that are designed to much lower live loads than what is needed for the present day normal traffic, it is first necessary to determine which bridges have an insufficient load-carrying capacity. In so doing, it is necessary to discover bridges' often hidden carrying capacities through analysis with modern and sophisticated methods. Such computerized tools of analysis have been established in Japan, are under consideration in Germany and Sweden, and are under development in Denmark.

Detailed information is the first requirement. When calculations and especially all "as-made" drawings are available, an inspection of the bridge in question is needed to confirm that no significant changes have been made that impede the reliability of the drawings.

In the United States the "AASHTO Manual for Maintenance Inspection of Bridges - 1978" contains guidance and procedures for rating of bridges where no design calculations are available as does the United Kingdom technical memorandum no BE 3/73 from the Department of the Environment: "The Assessment of Highway Bridges for Construction and Use Vehicles". The latter is under review and is expected to be in Limit State Format with both Mandatory and Advisory sections.

When insufficient information is available for a bridge rating, a tedious geometrical survey must be undertaken and the apparent external material properties be assessed with the aid of laboratory tests. For steel structures such surveys will go a long way toward assessing the carrying capacity of a superstructure, but for reinforced concrete further investigations that include stripping of the significant cross sections are needed.

The determination of load carrying capacity of foundations is likely to remain shadowy because of the apparent difficulties in confirming what in fact is below ground level. An alternative to a thorough inspection is to attempt a load test for each bridge support combined with a geotechnical investigation. By measuring the settlements for various loads and relating these facts with the geotechnical information, it may be possible to arrive at useful conclusions. If the year of construction is known, some knowledge as to the usual practice at that time, both with respect to design and construction, may prove helpful.

IV.4.2 Revised design criteria

The construction of road systems in the majority of OECD countries is basically completed and a major increase in loads would lower the useful life for a large number of bridges if heavier traffic loads were permitted. It is therefore generally agreed that the present design loads - although they vary from country to country - are sufficient and that only limited changes are warranted.

Some of the suggested changes are:

- increased load on superstructures and especially pedestrian bridges to counteract impact from vehicles,
- the establishment of a fatigue load-scheme for fatigue prone details; the need for such a load-scheme in rating procedures becomes evident when considering the many older bridges with rapidly increasing impacts from heavy freight vehicles;
- the development of ship collision design criteria that balance risks against costs.

A more permanent change in the external economic environment may reduce the overall expenditures to bridge building and bridge maintenance. In some countries the discount rate for these investments and expenditures have risen to a level of 9 per cent, net of inflation, causing concern as to the long term effects of such financial constraints. With a discount rate of 9 per cent the most economic life span of bridges may be reduced to 25 - 30 years and this will have a vast impact on future bridge design and maintneance.

If bridges are to be designed for such relatively short life spans the use of reinforced or prestressed concrete as well as in steel may be replaced by wood. The use of waterproofing and asphalt will almost certainly be abandoned. The frequency of replacement will ultimately reach 4 per cent per year causing continuous disruption of traffic. None of these effects have, however, yet materialised because of the natural inertia in developed societies to such revolutionary changes.

In all the OECD reports on bridges, the importance of an international collaboration and standardisation has been underlined again and again. In an area such as the European continent where heavy traffic moves freely from one country to the next, the need for uniform design criteria is obvious and there is no disagreement on this point.

Work along these lines is underway in the Commite Euro-International du Beton (CEB) and in the European Community. In the latter organisation, work with the purpose of harmonizing axle loads and gross vehicle weight in traffic codes has been going on for the last 10-15 years but unfortunately without the prospect of reaching any conclusions in the foreseeable future.

When traffic becomes more and more internationalised so must traffic codes, and therefore load assumptions in design codes for bridges must be internationally standardised too. The technical difficulties in the way of standardisation are as numerous as the countries involved, but the CEB Model Code has proven to be one of the major steps in the right direction.

IV.4.3 Adaptability to rehabilitation, strengthening and other improvements

Of the many existing bridge types some are more adaptable for intervention than others. Often insignificant details may contain the key to whether the alteration can be carried out quickly and inexpensively or turn into a major operation. Some examples will illustrate this.

When bridge elements cannot within reasonable costs be made as durable as required, they must be made easily replaceable. If parapets or edge beams are likely to have a limited life span due to deteriorating attacks from climate and de-icing salt, it is advisable that such elements do not form part of the main static structural system but can be replaced without interfering with this system, e.g. preferably as precast elements on concrete bridges.

As mentioned in the OECD report on bridge maintenance, bridge bearings and expansion joints must be arranged to secure easy access for inspection and maintenance and for replacements.

Designs for modern cantilever constructions do not normally include provisions for major rehabilitation works. Similar lack of consideration is found in a number of modern bridge designs where the structural layout involves complex static integration of a majority of bridge parts, where even limited rehabilitation work may interfere dangerously with the static stability.

It may therefore be advisable that a bridge design contain a thorough analysis of possible rehabilitation and strengthening procedures including estimated costs for such procedures before the original bridge design is accepted. If bridge designs include such analyses, the future would contain fewer costly surprises.

Chapter V

RECOMMENDATIONS FOR RESEARCH

This Chapter should be read in conjunction with the recommendations for research given in previous OECD reports on Inspection (1976), Evaluation of Load Carrying Capacity (1979) and Maintenance (1981). It is a truism that most of the technical background to these topics is fully relevant to rehabilitation and strengthening. Nevertheless it is necessary to restate this and to list related recommendations for research.

Although some of the topics are already being researched wholly or in part, there is still a need for additional effort. This is particularly important in relation to situations that differ from one location to another, for example traffic or wind loading.

In seeking to devise improved ways to rehabilitate and strengthen bridges it is necessary to meet criteria of practicality:

- techniques must be robust and should not be unduly sensitive to workmanship or defects;
- techniques must be resistant to environmental conditions;
- it is necessary for the bridgework to be done quickly and with minimum delays to traffic;
- techniques must be relatively cheap;
- techniques should preferably involve proven technology and new ideas should be introduced with due care and attention.

The needs for further research have been grouped under the following headings: assessment leading to a decision on which strategy to take, replacement of components, strengthening and repairs.

V.1 ASSESSMENT

Before devising a strategy of repair, strengthening or replacement, it is necessary to assess the structure in order to evaluate its load carrying capacity and potential for further service (remnant life).

V.1.1 Traffic loading characteristics

There is a need for cheaper and quicker methods of measuring axle loads and gross vehicle loads. It is necessary for the weighing unit to be easily fitted into the pavement surface and to operate without interrupting the traffic. In many instances it is required that the measurements be made simultaneously in two or more traffic lanes but this is expensive with currently available techniques. There are several types of dynamic weighing systems and these vary in cost and accuracy; it is necessary to define the required accuracy in relation to the end product of the work. Most systems weigh axle loads but can also give vehicle loads and other information. In addition to load effects, trafficking may also cause mechanical vibration which can lead to problems. Improved methods of data handling would enable records to be obtained for longer periods of time which are more statistically significant. In most assessments maximum likely loads are needed but numbers of occurrences, average occurrences and combinations may be equally important.

V.1.2 Wind loading characteristics

The above comments on requirements for measurements, recording and processing of traffic load data are equally applicable to wind loading. Inadequate attention has been given to devising equipment to activate data collection automatically when predetermined conditions occur. However, some work has been done in the United Kingdom on measurements of wind characteristics (speeds and directions) and response of long-span steel box girder cable-stayed bridges at Wye and Erskine(6)(*) and bridges at Cleddau and Kessock. In the United States measurements on a long-term basis are being made on the cable-stayed Luling Bridge near New Orleans.

V.1.3 Load testing

Surprisingly few load tests are carried out on structures in use mainly because of expense and the resulting interruption to traffic. Some very useful load testing and interpretive work has been done by the Ontario Ministry of Transportation(1) and the state-of-the-art has been reviewed(2). An attractive alternative to static loading is the measurement of bridge response under normal trafficking. This can be even more useful if known loads are driven across the bridge at intervals to give "bench marks". These techniques are still in the development stage and work is required to define the range of applicability and to give guidance on details such as types and positions of measurements and methods of subsequent analysis.

V.1.4 Remnant life

Although much is known about the fatigue performance under laboratory conditions of welded connections, reinforcement and concrete, there is inadequate knowledge about behaviour in the field, particularly of concrete. Also there is a need for data related to time or cycle dependent degradation of other components. The performance of details such as expansion joints is generally believed to be influenced by numbers of axle loads and bridge movements but more data are required to aid assessment.

It is necessary to distinguish between the design life and the real expectation of the life of a bridge. The two can differ substantially because it is necessary for design life to take account of worst possible cases and these are not necessarily applicable to the bridge in question. One such example of the demanding requirements of design codes is fatigue loading and the calculation of stress spectra. Because of the uncertainties involved, it is necessary to take worst cases so that the resulting stresses may be significantly higher than in practice. It is necessary to close this gap by measurements of dynamic stresses; a project is currently being sponsored for measurements on steel bridges by the European Coal and Steel Community(3).

*) See list of references at the end of this Chapter.

Remnant life can be influenced by several mechanisms, for example concrete cracked by fatigue loading(9) may in the long run be damaged more by corrosion than by the development of fatigue processes.

There is a need to improve the methods of evaluating remnant life on all fronts.

V.1.5 Condition surveys

There is a need for improved methods to measure actual states of stress, loss of prestress, size of defects and material properties. Non-destructive testing is required to be done at two levels: (a) routine - used by semi-skilled workers so that equipment must be cheap, simple, robust, easy to use and giving a wholly unambiguous output, (b) special surveys - used by skilled workers for very important appraisals such as long-span bridges or sensitive bridges on heavily trafficked routes. In the former case it may be justifiable to develop special equipment particularly where there are large numbers of the details to be examined, for example bearings on multi-span viaducts, hanger-cables on suspension bridges or welded connections on long-span steel bridges.

It is considered particularly important to develop existing inspection techniques for prestressed concrete structures so as to make them viable in the field. The need for in-depth condition surveys of prestressed concrete bridges without interfering with the structures will increase in the coming years and it is essential to develop practical tools for this.

There are several non-destructive techniques which are very sophisticated and can be made to work under laboratory conditions. These require further research to bring them to the stage where they can be used in the field either for routine surveys or special examinations.

Permanent surveillance by means of dedicated instrumentation offers a possible method to assist with condition surveys. There are methods available which automatically collect geometric data, such as inclinometers and strain gauges, but there is a need for techniques of using these data. Assessments should include consideration of the development of time dependent creep.

V.1.6 Analytical assessment

As with condition surveys there are two levels of analytical assessment:

a) Low level calculations based on assumptive criteria. Such analyses are usually elastic or, at most, pseudo-plastic. Because of these simplifications it is necessary to use relatively high factors of safety to make allowance for unexpected situations. Low level calculations can be made under any conditions provided the accuracy is acceptable.

b) High level calculations with minimal assumptions. These are truly non-linear and are usually made by the method of finite elements. High level calculations are mainly used in research and can only be made by specialist groups. High level calculations can sometimes result in bridges being shown capable of carrying bigger loads.

There is a need for development of analytical methods which have an accuracy and level of sophistication which is midway between high and low levels. It is necessary to develop programs of this level which can handle the special problems developed by bridges in the field. The types of problem which require analyses include:

- Masonry arch bridges having typical defects in geometry and material. This has received surprisingly little attention using modern analytical techniques.
- Steel bridges. There is a need for robust programs that can handle defects such as corroded members, damaged hanger-cables and incorrect structural action.
- Concrete bridges. Behaviour of concrete presents special problems for non-linear analyses which are currently being tackled. There is a need for programs that can handle a variety of defects including localised corrosion in the reinforcement, failed prestressing strands and defective concrete.

It is important to have the physical data necessary for the analytical calculations. This is especially so for data related to what actually pertains to the structure as opposed to notional conditions.

Codes for the design of new structures are generally unsuitable for the assessment of existing bridges. It is therefore necessary to develop methods and rules for assessment which take into account local conditions including precise measurements of movements and articulation.

V.2 REPAIRS

When repairing defective components or materials it is crucially important to work on a wholly correct diagnosis of the source of the trouble. <u>Repairs based on an incorrect strategy can worsen the situation.</u> Some of the materials used in rehabilitation work require strict adherence to requirements imposed by environmental conditions, for example resistance to temperature, moisture and combinations of them. There is a growing need for simple and robust repair techniques and equipment. There is also a need to be able to evaluate the performance of repairs which involve the mixing of new and old materials, possibly of different generic forms. Repair materials must be compatible with mature concrete and durable under all environmental conditions.

V.2.1 Defective concrete

Cracking or spalling of concrete can arise through a number of causes. Repairs may be required to restore structural integrity or merely seal the defective area to prevent ingress of corrosive materials such as de-icing salt. There are numerous types of repair material available but comparatively little data to aid selection of the best material for the job. Research is needed under relevant weathering conditions to produce information about long-term performances.

The consequences of cracks requires research to determine whether they are potentially harmful and when it is necessary to treat them. Injection and sealing of cracks may help matters, do very little or worsen the situation depending on the problem in question. At the moment there is insufficient information about the properties of sealants and their compatibility with mature concrete. In cases of corrosion there may be alternative and better methods of dealing with the concrete and this requires research.

A number of concrete structures has been built over the last decades without due consideration to durability. The ability to develop and produce concrete products of greater durability will be a prerequisite for the building, operation and maintenance of a reasonable network of roads with the use of only limited resources; this will be a general demand on all road and bridge engineers in the future. Continued research with a view to increasing durability is required in the fields of concrete composition, aggregate, the use of silica fumes and fly ash, the use of additives, cement composition and properties, concrete transportation, and construction techniques, etc. The study of the use of bridge waterproofing systems and the development of improved types of insulation with optimum properties and life should be continued. Better insight in the moisture-spreading mechanisms in concrete must be provided. It is not yet possible to surface-treat external concrete surfaces to the desired quality, and continued testing and development of products for surface treatment is needed. A practical measuring technique, capable of being used in the field for the measuring of the water-repellent properties on vertical as well as horizontal surfaces, has not yet been developed and there is a need for further research. Outward spreading moisture is decisive for the usability of a protective surface coating with a given degree of moisture proofing. It will be of great importance to study to what extent moisture will build-up under varying temperature and humidity conditions in protective surface coatings with different degrees of waterproofing.

V.2.2 Corroded reinforcement

In cases where reinforcement has corroded it is necessary to survey the structure and ensure that all defective material is located(7). Repairs can only be effected if all such material is removed but more research is required to determine how best to halt the process and make good. There are areas of uncertainty in relation to the choice of repair material and how to treat the interfaces. The electro-chemical conditions in concrete structures are only partially known and studies will have to be continued. Practical techniques for measuring potential differences must be developed so that they can be used by unskilled workers. Surface treatment of reinforcement and the effect of such treatment must be studied.

V.2.3 Fatigue repair

There have been reported cases of fatigue in welded connections of steel structures and in hanger-bars and hanger-cables of suspended structures. Repairs of steel components cannot be made by simply replacing the cracked component or by re-welding because such methods would be unlikely to achieve the same life as the first failure. It is necessary to couple the repair with some other action such as modifying the structural action, hole drilling at the end of the crack and introducing residual compressive stress. Although such techniques are known to improve fatigue lives there are insufficient quantitative data available for design purposes.

V.2.4 Bridge management

Owing to the extent of defects, a large number of structures should not be repaired, but should be demolished and replaced by new structures. The economic implications of various systems of bridge management are largely unknown and need more study. Society as a whole can save large amounts of money if the life of a defective structure can be utilised optimally, i.e. if limited maintenance is carried out and the replacement takes place as late as possible without incurring excessive risks to safety. This requires regular checking of the development of defects, the remaining load carrying capacity and other risks.

The checking of structures may, for example, be carried out by means of practical systems, designed to register and signal subsidence, horizontal or vertical movement, widening of cracks, changes in conditions of vibration etc. The further utilisation of the structure will depend on the reliability of the system. The development of applicable systems, capable of registering the required signals given by a structure in the process of distintegration is therefore required.

V.3 STRENGTHENING

Wholescale strengthening may be necessary in cases when bridges are found to be structurally inadequate for one reason or another.

V.3.1 Arch bridges

There are a variety of methods of strengthening arch bridges but they are generally carried out with inadequate supportive structural analysis. The need for improved methods of analysis is outlined in Section V.1. They are required not only for assessment of existing structures but also to enable the strengthening schemes to be optimised.

V.3.2 Old steel structures

The traditional methods of strengthening (introduction of extra load carrying elements or strengthening of existing ones) can be applied for old steel bridges provided that the type and characteristics of the existing steel material are exactly known. Non or slightly destructive tests should be developed to determine material characteristics.

V.3.3 Fatigue

With the increasing volumes and weights of vehicular traffic, fatigue of steel structures is becoming a problem which dictates the performance of a number of types of welded connection. The introduction of new design rules has enabled existing structures to be assessed and some to be shown to be at risk. The problems that arise can differ from the situation where it is necessary to repair fatigue cracking (see Section V.2.3). Strengthening to improve fatigue performance is not merely a question of reducing net sectional stresses and it is usually necessary to tackle the job by more than one design change. More information is required about factors that influence fatigue such as mode of structural action, type of connection and crack behaviour under different conditions.

V.3.4 Post-tensioning

It has been shown that corrosion of strands may occur in the vicinity of voids in post-tensioning ducts(4). The current state-of-the-art is inadequate in relation to (a) methods of detecting voids and (b) methods to deal with the voids when located. Research is required on both these topics and particularly to establish the long-term efficacy of schemes to fill the voids. Post-tensioning in itself can be used as a method of retrospectively raising the load carrying capacity of bridges (see Section III.5.7). There is a need to review and assess the engineering of schemes used to date in order to bring together such experience for future use. It would be very useful to have recommended design procedures for retrospective post-tensioning as a technique for strengthening inadequate bridges.

V.3.5 Bonded plating

Defective bridges can be strengthened externally by steel plating connected by resin bonding(5) (see Section III.5.6). Although a number of schemes have been carried out in different countries, the technique has developed on an ad hoc basis. There is as yet no accepted design procedures and, as with retrospective post-tensioning, there is a need to survey the engineering and performance of past schemes. Further research is required on long term performance under adverse conditions [techniques and materials used in aeronautics are generally irrelevant(8)].

V.3.6 Earthquakes

Design and construction of bridges in locations that could experience earth tremors presents special problems. Experience indicates that many bridges are at risk and there is a need to devise strengthening schemes. Approaches that require research include the development of methods of connecting retrospectively structural elements together to eliminate the risk of escalating mechanisms of collapse. Another approach that merits attention is by design modification to change the mode of dynamic response to one that is less damaging.

V.4 REPLACEMENT OF COMPONENTS

In many instances a structure can be strengthened or rehabilitated by the replacement of key components. Such work has of course, to be done with minimal delays to traffic and it is necessary to develop procedures which aid speed of construction and reduce costs.

V.4.1 Structural elements

Replacement of structural elements relates mainly to old bridges. The need for replacements can arise through types of deterioration such as severe local corrosion or damage caused by fire or vehicular impact. Alternatively replacement may be necessary if the original design of the component in question is found to be inadequate. Schemes for structural replacements are usually designed in isolation but it is feasible that savings in costs and improved engineering could be effected by studies of past successes and failures.

It is necessary to continue research on replacement of parts of prestressed concrete structures and of differential behaviour between new and old parts. The structural behaviour during repair also requires further research, for example behaviour of a partially reconstructed bridge under traffic loading and vibrations.

V.4.2 Decks

Common reasons for needing to replace concrete decks are through the development of defects due to defective concrete and through the spread of corrosion in the steel reinforcement. Replacements present special problems due to interruption to traffic and, more important, the weakening of the structure whilst the deck is being rebuilt. Corroded reinforced concrete decks can be patched but there are attendant risks that the process will continue. Moreover, the introduction of localised new material could worsen the situation so that corrosion is accelerated; work is required to assess this risk. In some cases of extreme corrosion it may be necessary to replace the whole deck. Research needs include appraisal of the different methods of repairing localised damage and the development of practical systems for rapid replacement of whole decks. Use of prefabricated sections is an attractive option but it is necessary to optimise methods of connection and coupling. There is also the question of what type of running surface to use.

Bridges found to have marginally inadequate load carrying capacity, such as old truss bridges, can sometimes be rehabilitated by the introduction of lightweight decks, for example steel orthotropic plates or lightweight aggregate concrete slabs. Both types of design present problems which require research; development of optimised design for the case in question, choice of the type of running surface and method of connection to existing beams. Further studies are required to determine the material performance of lightweight aggregate concrete.

V.4.3 Piers

It is rare for piers to have any corrosion protection and degradation can occur at caps due to leakage of chlorides from above and at road level due to splashing from passing vehicles. Research needs include the study of practical methods to replace piers using, say, prefabricated sections. Methods of doing local repairs using techniques such as injection of resin, application of protective coatings and cathodic protection require evaluation.

V.4.4 Cables

All forms of steel cable are liable to degradation through corrosion, corrosion fatigue, or stress-corrosion and may require replacement at one time or another. Immediate problems requiring research include diagnosis of the source of the trouble, method of supporting the bridge whilst the replacement work is being done and improved design to avoid or reduce live loads and give better corrosion protection. There is a need for better designs of sockets to reduce the high stresses experienced by outer wires and to increase the ease of articulation. It is becoming evident that cables fail after times that are relatively short in relation to notional design lives so that improvements based on the present state-of-the-art are unlikely to give adequate enhancements in durability. In consequence it is necessary to devise rehabilitation schemes that will permit easy replacements in future times. This applies to suspension bridges, cable-stayed bridges and post-tensioned bridges.

V.4.5 Expansion joints and bearings

Deterioration of expansion joints and bearings are common problems which can set up secondary structural actions which lead to more expensive troubles. There is a need for more research into factors that affect performance, including for example, detailed studies of relative movements between decks and bearings, so that improved components can be developed. There is also a need for designs that permit easier maintenance and replacement.

V.4.6 Waterproof membranes

Although many countries use waterproof membranes to protect bridge decks from corrosion, there is still some controversy about their efficacy and some prefer to protect the steel reinforcement with coatings. There is a need for continued surveillance to determine the performance of membranes and coatings in practice. This is needed because when a concrete deck is being rehabilitated it is often necessary to decide whether to use waterproofing in an effort to halt corrosion. There have been several reports on waterproofing (OECD, 1972; RILEM 1972; IABSE, Washington, September 1982; PIARC, Sydney, September 1983).

REFERENCES

1. BAKHT B and CSAGOLY PF. Diagnostic testing of a bridge. J. of Struc. Div. ASCE, 1980.

2. THOMPSON DM. Loading tests on highway bridges: a review. TRRL Report LR 1032. Crowthorne, 1981.

3. PAGE J and TILLY GP. Some analyses of traffic data for three steel bridges. TRRL Supplementary Report SR 598. Crowthorne, 1980.

4. WOODWARD RJ. Conditions within ducts in post-tensioned prestressed concrete bridges. TRRL Report LR 980. Crowthorne, 1981.

5. RAITHBY KD. External strengthening of concrete bridges with bonded steel plates. TRRL Supplementary Report SR 612. Crowthorne, 1980.

6. HAY JS. The wind induced response of the Erskine Bridge. Proc. Bridge Aerodynamics Conference, Inst. Civ. Engrs. London, March 1981.

7. VASSIE PR. A survey of site tests for the assessment of corrosion in reinforced concrete. TRRL Report LR 953. Crowthorne, 1980.

8. MAYS G and TILLY GP. Long endurance fatigue performance of bonded structural joints. Int. J. Adhesion and Adhesives Vol. 2 No. 2. April 1982.

9. SONODA K and HORIKAWA T. Fatigue strength of reinforced concrete slabs under moving loads. IABSE Colloquium Lausanne. IABSE Reports Vol. 37. Lausanne, 1982.

Chapter VI

CONCLUSIONS AND RECOMMENDATIONS

VI.1 THE NEED FOR A POLICY

The bridges on a highway network play a very specific role and their serviceability at all times, and acceptable conditions of safety, must be ensured.

The scale of the problem may be gauged from the simple fact that there are an estimated one million bridges with spans greater than 5 m in the OECD countries, of which almost half are on United States territory.

To ensure that service levels keep pace with needs, action is needed on two fronts apart from maintenance as defined in the OECD Report on Bridge Maintenance:

- rehabilitation to restore the original service level;
- strengthening to raise the service level to meet new demands.

Rehabilitation and strengthening works should be undertaken with a view to overall optimisation, particularly from an economic standpoint.

All the countries participating in the Group's study have established inspection procedures at three or four levels as recommended in the OECD Report on Bridge Inspection. The assessment of the data obtained by the help of the national inspection schemes indicates that unless an immediate start is made on a comprehensive rehabilitation programme of the existing bridge stock, considerable sums will have to be expended some ten years hence if the bridges are to be kept in use.

Given the importance of the stakes and the difficult economic situation, bridge rehabilitation and improvement have become matters of urgency calling for broad policy decisions encompassing all the other measures, notably maintenance, needed to ensure that bridges can operate in "acceptable" conditions.

VI.2 POLICY GUIDELINES

Broad policy guidelines for the rehabilitation and strengthening of existing bridges are discussed in detail in Chapter IV. They are summarised below.

VI.2.1 Needs

To work out a policy, one must first forecast needs. There are several aspects involved:

- _Traffic trends_ as regards traffic volume, vehicle loads and size. It is largely as a result of these trends that bridges require rehabilitation or strengthening.
- _Environmental constraints_. The increasing trend to favour preservation and safeguarding the environment when carrying out bridge works, in particular avoiding permanent changes (e.g. rechannelling of watercourses). There is an increasing need to install new equipment on existing bridges, such as noise barriers.
- _Socio-economic constraints_. These apply to increased demand for overall safety and to maintaining existing service levels; avoiding traffic restrictions during rehabilitation and strengthening may imply major additional outlays.

These demands are liable to increase substantially the cost of rehabilitation. Cost/benefit analyses are called for with a view to taking these demands into account.

VI.2.2 The tools

In planning rehabilitation and strengthening work, priority setting concepts using discount rates can be employed. This method is suggested in order to estimate future expenditure when evaluating the various possible solutions to a given problem.

The use of this tool may imply lower discount rates for important structural bridge elements than for bridge components with a shorter life; the discount rate for major bridges may also be lower than for ordinary bridges.

VI.2.3 Comparison between rehabilitation (or strengthening) and replacement

A choice must be made between rehabilitation or strengthening and replacement. Intermediate solutions such as limited rehabilitation, serving to postpone replacement, may also be envisaged.

Their respective merits should be analysed in the light of the following factors:

- the costs and benefits including those occurring in the future;
- the technical and financial uncertainties of rehabilitation or strengthening;
- inconvenience caused by replacement (to users, residents, possible environmental damage, etc.).

Such analyses should be made in all cases where works on any significant scale are needed on existing structures.

VI.2.4 Policy considerations

In conclusion, the considerations on which any policy must be based are:

- how bridges perform over time,
- how needs are likely to evolve,
- the state-of-the-art in regard to technology,
- the resources available.

These aspects are discussed more fully in the following paragraphs.

VI.3 PERFORMANCE OF BRIDGES OVER TIME

This issue was discussed in considerable depth in the OECD report on Bridge Maintenance to which the reader is referred. It is of crucial importance for rehabilitation and strengthening and deserves to be brought up again here so as to elucidate a number of aspects. It is necessary, in particular, to clarify thinking on replacement rates and lifespan and see what conclusions can be drawn from our knowledge of the existing bridge stock.

VI.3.1 Replacement rate

The rate of bridge replacement is closely linked to the average age of the existing stock and assumptions as to the desirable average lifespan. The general consensus among countries is that the target for the desired lifespan should be in the order of 100 years.

Given the very considerable investment over the last few decades, the proportion of bridges built in the last few decades is high. With the general slackening off in investment which is currently occurring, the average age of bridges in service is likely to rise significantly over the next few years. The rate of bridge replacement will therefore increase with a subsequent levelling off in about 30-40 years. This increase in the replacement rate is closely linked to the ability and will to maintain the present bridge stock.

Unless a very real effort to match the bridge stock to needs, the gap between them will widen and the replacement rate will be higher than the natural replacement rate. This will lead to a higher total expenditure.

VI.3.2 Knowledge of the existing bridge stock

The need to complete and store data on existing bridges has already been highlighted in the OECD report on Bridge Maintenance. Particular importance should be attached to the use as a management tool of a computerised general inventory in a policy of rehabilitation or strengthening.

United States experience illustrates the value of such an inventory in deciding how best to allocate funds for the rehabilitation (or replacement) of a large number of bridges.

A general data bank supplemented where needed by more detailed local records is considered essential to the decision-making process and to the optimisation of resources.

VI.3.3 Measures to be taken at the construction stage

In reviewing major rehabilitation or repair projects implemented in recent years, a number of conclusions emerge which it is recommended be taken into account in designing future structures:

Improvements are needed in:

- bridge design to create sturdier structures and make better allowance for likely future loads;
- making allowance for phenomena which up to now have been insufficiently catered for or even ignored (e.g. damage caused by shipping accidents);

- adapting bridges to facilitate future interventions; it is particularly important that bridge elements whose projected lifespans are far shorter than that assumed for the bridge as a whole should be easily replaceable. Their mode of replacement should be specified at the design stage and the way of doing so worked out in detail at that time.

VI.4 FUTURE NEEDS

VI.4.1 Trends in needs of normal traffic

Chapter IV provides a basis for gauging how ordinary traffic needs might evolve (excluding exceptional convoys).

Highway networks, and particularly the bridges along them, have been improved to meet constantly increasing traffic demands. Today they represent a very considerable capital investment and will in future largely condition how authorities approach the problem of coping with future needs.

In the last analysis, it is probable that total traffic volume will continue to grow and that the proportion of heavy vehicles in traffic as a whole will continue to increase for many years yet.

The provision and application of efficient control measures against overloading of normal freight vehicles are necessary.

VI.4.2 Exceptional heavy vehicles

This question has been treated in some depth in the OECD report on the Evaluation of Load Carrying Capacity of Bridges, to which the reader is referred. Given the trends which are today emerging in the different countries, the following recommendations would appear warranted:

- Most major roads could be improved to allow the passage of convoys of a total weight of up to 100 tonnes. For heavier convoys a network of special routes capable of carrying such abnormal loads should be provided. If bridges are subjected to very heavy loads for which they were not designed, the resulting repercussions for future expenditure on strengthening and rehabilitation will be considerable.
- Countries should work out broad policies for exceptional convoys in the light of existing transport infrastructure; these should not be confined to roads alone.
- Road transport policies generally should be harmonized between countries on a continent-wide basis; this harmonization should cover such aspects as load classification and the construction of vehicles.

VI.5 TECHNIQUES

Progress still needs to be made in some areas of rehabilitation and strengthening techniques. Chapter V sets out the areas where a particular research effort is needed and the reader is referred to this Chapter.

In deciding on the approach to be taken on particular cases as well as the techniques to be applied, account must be taken of the present state-of-the-art and the scope and limitations of available techniques.

A comprehensive policy of rehabilitation and strengthening must take account of any progress in these techniques.

VI.6 RESOURCES

VI.6.1 Financial resources

From the evaluation made in the various countries, it emerges that the funds allocated to bridge rehabilitation or renovation fall very often substantially short of needs.

The OECD report on Bridge Maintenance puts the annual financial resources required for maintenance at 0.5 per cent of the replacement value. To maintain service levels for traffic needs and making the best use of existing bridges, additional appropriations are needed for major repairs and the replacement of wornout structures. These appropriations should not be less than 1.5 per cent of the replacement value of the total bridge stock, excluding expenditure on strengthening or geometric improvements which constitutes capital investment. This ratio is a ceiling which assumes a lifespan of around 80-100 years. It may be substantially exceeded, depending on local conditions. In Denmark, for instance, where climatic conditions are generally severe, the forecast for the year 2000 is 2.5 per cent as against 1.3 per cent for 1980; in the United States the ratio at present is over 3 per cent reflecting a very considerable modernisation and, hence, investment effort.

VI.6.2 Manpower resources

Staff

The slackening off in the rate of investment would appear to be accompanied in some countries by a comparable fall in manpower allocated to bridges. The establishment of a policy of rehabilitation and strengthening on an ongoing basis should serve to exploit the skills of staff already conversant with bridge techniques.

In the years to come, as many people will be required for the technical tasks implicit in bridge management as for those of bridge construction in the past.

Technical training

Repair and strengthening may involve recourse to sophisticated techniques requiring special training. Such training may be acquired more easily by staff with prior experience in new construction.

Inasmuch as these are tasks which will have to be performed for a very long time, it is imperative to train the personnel now.

Engineering instruction, which at present is highly biased towards new construction, should thus include specific training in the technical tasks of bridge management (inspection, maintenance, rehabilitation).

A factor in the effectiveness of training in this area, whether in normal technical education establishments or in the form of retraining courses, is how the full use of data collected on the occasion of particular bridge repair operations can be exploited. Arrangements for gathering and disseminating such information should be set up at central level, a process already under way in some countries.

This information is needed as much by staff of the bridge administration as by personnel working for contractors.

Lastly, special training should be given to staff of local and regional enterprises who represent a major potential but who are often unable to cope with the technical problems posed by repairs.

VI.7 GENERAL CONCLUSIONS

To recapitulate, the main recommendations put forward in this Chapter are the following:

- there is an urgent need for comprehensive policies encompassing maintenance, rehabilitation, strengthening or improvement, and replacement;
- in formulating such policies a different weighting should be attributed to future outlays depending on whether these concern structural components with longer or shorter lifespans on major or other bridges;
- such policies should be based on forecasts of future needs: likely traffic trends, environmental or socio-economic constraints;
- forecasts of traffic trends should assume a likely increase both in total volume and in the proportion of heavy traffic; forecasts for exceptional convoys should be made and the traffic rules for such convoys should be worked out on a continent-wide basis;
- replacement should be considered as an alternative solution to rehabilitation or strengthening;
- the design of new bridges should allow for subsequent needs and the bridges should be built so as to facilitate future repair;
- data on the existing bridge stock should be compiled and stored so that they can be exploited for policy purposes;
- progress is still needed on certain specific repair techniques; research in this area should be expanded and pursued as a matter of urgency;
- an annual sum equivalent to 1.5 - 2 per cent of the replacement value of the total bridge stock should be earmarked for rehabilitation or the replacement of "below standard" bridges;
- a particular effort should be devoted to staff training; in the short term this may be confined to manpower already conversant with the techniques of new construction. With a view to a longer-term solution, training further upstream should be envisaged for future staff.

Inasmuch as the problem is posed in similar terms in all the participating countries and is both recent and complex, it is strongly urged that the international co-operation which enabled this report to be **produced** be pursued in its application.

ANNEX

RESULTS OF ENQUIRY ON SUBJECTS COVERED IN CHAPTER IV

Questions	Belgium	Denmark	Finland	France
1. Considerations regarding life span of bridges.	50-100 years	Life span not a design criterion. Expected average life of bridges 50-70 years.	Present average 35 years. Target average 70 years in year 1990. Ordinary timber 25 years. Glued beams timber 40 years. Reinforced concrete 70 years. Steel 90 years. Major bridges 100 years.	Between 100 years and several centuries.
2. Advice for design criteria for future traffic needs.	Ongoing revision of present codes for live loads, including provisions for EHV's.	Evaluation of ship collisions with respect to risks and costs.	Present design criteria regarded as sufficient for future needs.	Present design criteria regarded as sufficient for future needs.
3. Solutions for seepage of water through bridge decks.	Waterproofing required. Various kinds used according to type of bridge.	Waterproofing mandatory. Now only use of prefab. reinforced bituminous sheets.	Waterproofing required. Various kinds used according to type of bridge.	Waterproofing required. Several kinds used.
4. Ongoing activity concerning improvement of administration of exceptionally heavy vehicles.	Ongoing standardization of EHV's. Heavy load grid established.	Heavy load grid established. Ongoing research for computerising administration.	Increased law enforcement against overload violations.	Many existing rules. Extensive study finalised but not yet published.
5. Existing rules or codes for rating bridges without available information such as drawings and calculations.	No existing rules or codes. Normal bridge loading should be used.	No existing rules or codes.	No existing rules or codes.	No existing rules or codes. Problems are solved ad hoc.
6. Any ongoing research for establishing rating procedures for such bridges.	No ongoing research.	Research project under way.	No ongoing research.	No ongoing research.
7. Number of bridges not yet rated and the importance hereof.	Only minor and old bridges.	6-8,000 bridges. The problem is considered important.	Only few and minor bridges. The problem is not considered important.	Only few bridges not rated. Problem considered important.

Germany	Italy	Japan	Netherlands	Norway
The "Ablösungsrichtlinien" are regarded reliable for average life spans. Great individual deviations are underlined. Sophisticated research ongoing.	Some ongoing studies. No established consideration.	Design assumptions for ordinary bridges are 50 years. For major bridges 100 to 200 years.	Expected life of about 100 years which is confirmed considering existing bridges. Ongoing research on the subject.	50-100 years is regarded as reasonable.
- Fatigue load scheme - Higher multiple axle load - Simultaneous heavy load - Reduced dead load under repair but with traffic.	Fatigue load scheme based on limit state.	Present design criteria regarded as sufficient for future needs.	Ongoing work for revision of existing criteria. No great changes are expected.	Present design criteria regarded sufficient for future needs.
Waterproofing mandatory.	Increasing use of waterproofing. Ongoing research are types to be used.	Generally no action needed. In rare cases waterproofing is used.	For steel bridges waterproofin is used. On concrete bridges a bituminous bonding layer under an asphaltic concrete layer is used.	Operates with 6 classes of bridge deck treatments, which are applied according to traffic volume and amount of de-icing salt used.
Continuous improvements in rating and administrative procedures. Call for better international co-operation.	No ongoing activities.	EDP-programmes available for the administration when analysing request for permit for EHV.	An extensive study has been	New and revised procedures for permits for exceptional vehicles until 80 tons into effect during 1983. For heavier transports, nothing planned so far.
No existing rules or codes. Problems solved in classical manner.	Normal engineering approach based on a survey of the structure in question.	No existing rules or codes.	No existing rules or codes.	Existing guidelines available for rating procedures.
No ongoing research.	Research studies about to commence.	No ongoing research work.	No ongoing research work.	No ongoing research.
10-30 per cent of bridges on rural roads. The problem is not considered serious.	About 10 per cent of all structures. Problem considered important.	- Considered important. - No accurate estimate available, but roughly assessed to 20 per cent of total number.	- Considered important. - No inventory made.	Only a small number of minor bridges. The problem is not considered important.

Spain	Sweden	Switzerland	United Kingdom	United States
No attempt to qualify life span.	Aiming at a life of 100 years when erecting new bridges.	Approximately 80-100 years is regarded as reasonable.	Design assumptions established assuming 120 years life for the structure as such.	Design for 50-60 years for ordinary bridges and 100 years for major bridges.
Present design criteria regarded as sufficient for future needs.	Present design criteria insufficient for the secondary static system in superstructures only.	Expected increase in total weight of vehicles to conform with European standard.	Present design criteria give some scope for changes in traffic patterns	HS 25 seems appropriate. Large number of existing bridges with reduced life when heavier loads permitted.
Generally no action needed. Recommendation to use waterproofing in exposed areas.	Waterproofing is always required. Good experience with mastic asphalt. Prefab. reinforced bituminous sheets may be more widely used.	Waterproofing required. Importance of detailing underlined.	Use of waterproof membranes on a prequalification basis.	Epoxy coating of rebars plus concrete overlay - either high density or latex modified. Existing bridges may apply cathodic protection.
Itineraries established and expanded as the need arises.	Have recently established special road network for EHV. A speical EDP-based analysis system under consideration.	Itineraries established for various levels of loads.	Heavy load grid established 1970 and operates satisfactorily. Movements of heavy loads are strictly controlled.	Considerable ongoing activities by researchers and several committees.
No existing rules or codes.	No existing rules or codes, but a general trend is developed through experience which include test loading.	No existing rules or codes. Problems are solved ad hoc.	DTp Technical Memorandum BE 3/73	AASHTO Manual for Maintenance Inspection of Bridges - 1978.
Research projects under consideration.	No ongoing research.	A research study about to commence.	BE 3/73 currently under revision. Some ongoing research	Several research projects under way.
- Regarded as very important. - No estimation as to number of bridges not rated.	About 3,000 on rural roads but the problem is under control and is not regarded as important.	Problem considered important.	Regarded as a potential problem.	60,000 bridges outside the federal-aided highway system. The problem considered fairly important.

LIST OF MEMBERS OF THE GROUP

Chairman: Mr. C. Bois, France

BELGIUM

Mr. J. de BUCK
Ingénieur en Chef-Directeur
Ministère des Travaux publics
Bureau des Ponts
Direction générale - 1ère Division
rue Guimard, 9
1040 BRUXELLES

Mr. V. VEVERKA
Chef du Service "Conception et Structures"
Centre de Recherches Routières
Boulevard de la Woluwe, 42
1200 BRUXELLES

DENMARK

Mr. Per CLAUSEN
Head of Maintenance Section

Mr. H.-H. GOTFREDSEN
Head of Bridge Division

The Highway Directorate
Havnegade 23
P.O. Box 2169
1016 COPENHAGEN

FINLAND

Mr. Kalevi FALCK, Civ. Eng.
Roads and Waterways Administration
Bridge Construction Division
Box 33
00521 HELSINKI 52

FRANCE

M. GRELU
Ingénieur en Chef des Ponts et Chaussées
Directeur de la Division Etudes Générales et
 Ouvrages-Types du Département Ouvrages d'Art
 du SETRA
46, avenue Aristide Briand, B.P. 100
92223 BAGNEUX CEDEX

M. C. BOIS
Laboratoire Central des Ponts et Chaussées (LCPC)
58 Bd. Lefebvre
75732 PARIS CEDEX 15

GERMANY

Mr. D.E. LEBEK
Bundesanstalt für Strassenwesen
Brühlerstrasse 1
5000 KOLN 51

ITALY
Sig. Ing. Enzo FREDIANI
Soc. AUTOSTRADE
Via Nibby 10
00161 ROME

JAPAN
Mr. Shoichi SAEKI
Chief, Bridge Research Section
Public Works Research Institute
Ministry of Construction
Asahi 1-chome, Toyosato-machi
Tsukuba-gun
IBARAKI-KEN 305

THE NETHERLANDS
Mr. M.F.A. EL-MARASY
Bridges Directorate
Rijkswaterstaat
Postbus 285
2270 AG VOORBURG

NORWAY
Mr. Olav GRINDLAND
Norwegian Public Roads Administration
Bridge Division, P.O. Box 8109, Dep
OSLO 1

SPAIN
Mr. Ramón del CUVILLO
Ingénieur en Chef de la Section des Ponts
Direction Général des Routes
Ministerio de Obras Públicas
MADRID - 3

D. L. ORTEGA BASAGOITI
Laboratorio Central de Estructuras y Materiales
C/Alfonso XII, no 3
MADRID 7

SWEDEN
Mr. Lennart LINDBLADH
National Swedish Road Administration
Bridge Section
781 87 BORLANGE

SWITZERLAND
Prof. J.-C. BADOUX
Directeur
ICOM-Construction métallique
Ecole polytechnique fédérale
GCB (Ecublens)
CH-1015 LAUSANNE

Mr. P.A. MATTHEY
Assistant à l'Ecole polytechnique fédérale
ICOM-Construction métallique
GCB (Ecublens)
CH-1015 LAUSANNE

Mr. P. SCHMALZ
Adjoint scientifique
Office fédéral des routes
Monbijoustrasse 40
3003 BERN

TURKEY	Mr. Tomris YARDIMCI (Corresponding Member) Deputy Director of Bridge Maintenance Division Karayollari Genel Müdürlüğü Köprüler Dairesi Baskanligi ANKARA
UNITED KINGDOM	Dr. G.P. TILLY Head, Bridges Division Highways and Structures Department Transport & Road Research Laboratory (TRRL) Old Wokingham Road CROWTHORNE, Berks RG11 6AU
UNITED STATES	Mr. Charles F. GALAMBOS Chief, Structures Division Office of Engineering & Highway Operations Research and Development Mr. Stanley GORDON Bridge Division Office of Engineering Federal Highway Administration Department of Transportation WASHINGTON, D.C. 20590
CEB	Mr. R. TEWES EPFL Ecublens CEB, CP 88 1015 LAUSANNE, Suisse
OECD	Mr. B. HORN
Rapporteurs	Messrs. Bois, Grelu, Lebek, Gotfredsen and Tilly
Members of the Editing Committee	Messrs. Bois, de Buck, Gotfredsen, Grelu, Horn, Lebek and Tilly

OECD SALES AGENTS
DÉPOSITAIRES DES PUBLICATIONS DE L'OCDE

ARGENTINA – ARGENTINE
Carlos Hirsch S.R.L., Florida 165, 4° Piso (Galería Guemes)
1333 BUENOS AIRES, Tel. 33.1787.2391 y 30.7122

AUSTRALIA – AUSTRALIE
Australia and New Zealand Book Company Pty, Ltd.,
10 Aquatic Drive, Frenchs Forest, N.S.W. 2086
P.O. Box 459, BROOKVALE, N.S.W. 2100

AUSTRIA – AUTRICHE
OECD Publications and Information Center
4 Simrockstrasse 5300 BONN. Tel. (0228) 21.60.45
Local Agent/Agent local :
Gerold and Co., Graben 31, WIEN 1. Tel. 52.22.35

BELGIUM – BELGIQUE
Jean De Lannoy, Service Publications OCDE
avenue du Roi 202, B-1060 BRUXELLES. Tel. 02/538.51.69

BRAZIL – BRÉSIL
Mestre Jou S.A., Rua Guaipa 518,
Caixa Postal 24090, 05089 SAO PAULO 10. Tel. 261.1920
Rua Senador Dantas 19 s/205-6, RIO DE JANEIRO GB.
Tel. 232.07.32

CANADA
Renouf Publishing Company Limited,
2182 ouest, rue Ste-Catherine,
MONTRÉAL, Qué. H3H 1M7. Tel. (514)937.3519
OTTAWA, Ont. K1P 5A6, 61 Sparks Street

DENMARK – DANEMARK
Munksgaard Export and Subscription Service
35, Nørre Søgade
DK 1370 KØBENHAVN K. Tel. +45.1.12.85.70

FINLAND – FINLANDE
Akateeminen Kirjakauppa
Keskuskatu 1, 00100 HELSINKI 10. Tel. 65.11.22

FRANCE
Bureau des Publications de l'OCDE,
2 rue André-Pascal, 75775 PARIS CEDEX 16. Tel. (1) 524.81.67
Principal correspondant :
13602 AIX-EN-PROVENCE : Librairie de l'Université.
Tel. 26.18.08

GERMANY – ALLEMAGNE
OECD Publications and Information Center
4 Simrockstrasse 5300 BONN Tel. (0228) 21.60.45

GREECE – GRÈCE
Librairie Kauffmann, 28 rue du Stade,
ATHÈNES 132. Tel. 322.21.60

HONG-KONG
Government Information Services,
Publications/Sales Section, Baskerville House,
2/F., 22 Ice House Street

ICELAND – ISLANDE
Snaebjörn Jönsson and Co., h.f.,
Hafnarstraeti 4 and 9, P.O.B. 1131, REYKJAVIK.
Tel. 13133/14281/11936

INDIA – INDE
Oxford Book and Stationery Co. :
NEW DELHI-1, Scindia House. Tel. 45896
CALCUTTA 700016, 17 Park Street. Tel. 240832

INDONESIA – INDONÉSIE
PDIN-LIPI, P.O. Box 3065/JKT., JAKARTA, Tel. 583467

IRELAND – IRLANDE
TDC Publishers – Library Suppliers
12 North Frederick Street, DUBLIN 1 Tel. 744835-749677

ITALY – ITALIE
Libreria Commissionaria Sansoni :
Via Lamarmora 45, 50121 FIRENZE. Tel. 579751/584468
Via Bartolini 29, 20155 MILANO. Tel. 365083
Sub-depositari :
Ugo Tassi
Via A. Farnese 28, 00192 ROMA. Tel. 310590
Editrice e Libreria Herder,
Piazza Montecitorio 120, 00186 ROMA. Tel. 6794628
Costantino Ercolano, Via Generale Orsini 46, 80132 NAPOLI. Tel. 405210
Libreria Hoepli, Via Hoepli 5, 20121 MILANO. Tel. 865446
Libreria Scientifica, Dott. Lucio de Biasio "Aeiou"
Via Meravigli 16, 20123 MILANO Tel. 807679
Libreria Zanichelli
Piazza Galvani 1/A, 40124 Bologna Tel. 237389
Libreria Lattes, Via Garibaldi 3, 10122 TORINO. Tel. 519274
La diffusione delle edizioni OCSE è inoltre assicurata dalle migliori librerie nelle città più importanti.

JAPAN – JAPON
OECD Publications and Information Center,
Landic Akasaka Bldg., 2-3-4 Akasaka,
Minato-ku, TOKYO 107 Tel. 586.2016

KOREA – CORÉE
Pan Korea Book Corporation,
P.O. Box n° 101 Kwangwhamun, SÉOUL. Tel. 72.7369

LEBANON – LIBAN
Documenta Scientifica/Redico,
Edison Building, Bliss Street, P.O. Box 5641, BEIRUT.
Tel. 354429 – 344325

MALAYSIA – MALAISIE
University of Malaya Co-operative Bookshop Ltd.
P.O. Box 1127, Jalan Pantai Baru
KUALA LUMPUR. Tel. 51425, 54058, 54361

THE NETHERLANDS – PAYS-BAS
Staatsuitgeverij, Verzendboekhandel,
Chr. Plantijnstraat 1 Postbus 20014
2500 EA S-GRAVENHAGE. Tel. nr. 070.789911
Voor bestellingen: Tel. 070.789208

NEW ZEALAND – NOUVELLE-ZÉLANDE
Publications Section,
Government Printing Office Bookshops:
AUCKLAND: Retail Bookshop: 25 Rutland Street,
Mail Orders: 85 Beach Road, Private Bag C.P.O.
HAMILTON: Retail Ward Street,
Mail Orders, P.O. Box 857
WELLINGTON: Retail: Mulgrave Street (Head Office),
Cubacade World Trade Centre
Mail Orders: Private Bag
CHRISTCHURCH: Retail: 159 Hereford Street,
Mail Orders: Private Bag
DUNEDIN: Retail: Princes Street
Mail Order: P.O. Box 1104

NORWAY – NORVÈGE
J.G. TANUM A/S Karl Johansgate 43
P.O. Box 1177 Sentrum OSLO 1. Tel. (02) 80.12.60

PAKISTAN
Mirza Book Agency, 65 Shahrah Quaid-E-Azam, LAHORE 3.
Tel. 66839

PHILIPPINES
National Book Store, Inc.
Library Services Division, P.O. Box 1934, MANILA.
Tel. Nos. 49.43.06 to 09, 40.53.45, 49.45.12

PORTUGAL
Livraria Portugal, Rua do Carmo 70-74,
1117 LISBOA CODEX. Tel. 360582/3

SINGAPORE – SINGAPOUR
Information Publications Pte Ltd,
Pei-Fu Industrial Building,
24 New Industrial Road N° 02-06
SINGAPORE 1953, Tel. 2831786, 2831798

SPAIN – ESPAGNE
Mundi-Prensa Libros, S.A.
Castelló 37, Apartado 1223, MADRID-1. Tel. 275.46.55
Libreria Bosch, Ronda Universidad 11, BARCELONA 7.
Tel. 317.53.08, 317.53.58

SWEDEN – SUÈDE
AB CE Fritzes Kungl Hovbokhandel,
Box 16 356, S 103 27 STH, Regeringsgatan 12,
DS STOCKHOLM. Tel. 08/23.89.00
Subscription Agency/Abonnements:
Wennergren-Williams AB,
Box 13004, S104 25 STOCKHOLM.
Tel. 08/54.12.00

SWITZERLAND – SUISSE
OECD Publications and Information Center
4 Simrockstrasse 5300 BONN. Tel. (0228) 21.60.45
Local Agents/Agents locaux
Librairie Payot, 6 rue Grenus, 1211 GENÈVE 11. Tel. 022.31.89.50

TAIWAN – FORMOSE
Good Faith Worldwide Int'l Co., Ltd.
9th floor, No. 118, Sec. 2,
Chung Hsiao E. Road
TAIPEI. Tel. 391.7396/391.7397

THAILAND – THAILANDE
Suksit Siam Co., Ltd., 1715 Rama IV Rd,
Samyan, BANGKOK 5. Tel. 2511630

TURKEY – TURQUIE
Kultur Yayinlari Is-Türk Ltd. Sti.
Atatürk Bulvari No : 77/B
KIZILAY/ANKARA. Tel. 17 02 66
Dolmabahce Cad. No : 29
BESIKTAS/ISTANBUL. Tel. 60 71 88

UNITED KINGDOM – ROYAUME-UNI
H.M. Stationery Office, P.O.B. 276,
LONDON SW8 5DT. Tel. (01) 622.3316. or
49 High Holborn, LONDON WC1V 6 HB (personal callers)
Branches at: EDINBURGH, BIRMINGHAM, BRISTOL,
MANCHESTER, BELFAST.

UNITED STATES OF AMERICA – ÉTATS-UNIS
OECD Publications and Information Center, Suite 1207,
1750 Pennsylvania Ave., N.W. WASHINGTON, D.C.20006 – 4582
Tel. (202) 724.1857

VENEZUELA
Libreria del Este, Avda. F. Miranda 52, Edificio Galipan,
CARACAS 106. Tel. 32.23.01/33.26.04/31.58.38

YUGOSLAVIA – YOUGOSLAVIE
Jugoslovenska Knjiga, Knez Mihajlova 2, P.O.B. 36, BEOGRAD.
Tel. 621.992

Les commandes provenant de pays où l'OCDE n'a pas encore désigné de dépositaire peuvent être adressées à :
OCDE, Bureau des Publications, 2, rue André-Pascal, 75775 PARIS CEDEX 16.
Orders and inquiries from countries where sales agents have not yet been appointed may be sent to:
OECD, Publications Office, 2, rue André-Pascal, 75775 PARIS CEDEX 16.

OECD PUBLICATIONS, 2, rue André-Pascal, 75775 PARIS CEDEX 16 - No. 42789 1983
PRINTED IN FRANCE
(77 83 04 1) ISBN 92-64-12528-0